Lecture Notes in Control and Information Sciences

Edited by M. Thoma and A. Wyner

For information about Vols. 1–80 please contact your bookseller or Springer-Verlag.

Lecture Notes in Control and Information Sciences

Edited by M. Thoma and A. Wyner

145

M. B. Subrahmanyam

Optimal Control with a Worst-Case Performance Criterion and Applications

Springer-Verlag Berlin Heidelberg GmbH

Author

M. Bala Subrahmanyam
Flight Control, Code 6012
Air Vehicle & Crew Systems Technology Dept.
Naval Air Development Center
Warminster, PA 18974-5000
USA

ISBN 978-3-540-52822-7   ISBN 978-3-540-47158-5 (eBook)
DOI 10.1007/978-3-540-47158-5

This monograph is lovingly dedicated to

JESUS CHRIST,

God's gift of love.

[1 JOHN 4:10]

PREFACE

This monograph deals with optimal control problems in which the cost functional is a product of powers of definite integrals. In particular, we consider cost functionals of the form of a quotient of definite integrals and their relation to finite-interval H_∞ control, performance robustness, and model reduction.

The material in the book is taken from a collection of research papers written by the author. These research papers are reproduced here without much alteration. Thus there is some duplication of material in the various chapters and the chapters are reasonably independent of one another. As far as further research is concerned, I feel that there is a need in the areas of performance robustness and computational algorithms.

In our compuations, we found the function which represents the worst-case performance measure (denoted by λ in the various chapters) to be a difficult function to evaluate. A vast amount of computer time was spent in evaluating this function for given control parameters. Further research needs to be done to find an alternate method to evaluate λ more accurately and efficiently. We found it especially difficult to evaluate the global maximum of this function because of the existence of a large number of local maxima. However, in most cases it is adequate to obtain a good local maximum. The area of model reduction is a notable exception. In this case one needs to look for the global maximum. Also, inaccurate evaluation of λ may prematurely terminate the optimization procedure which seeks to maximize λ. In any case, it is hoped that this monograph will stimulate additional work in the area of evaluation of λ and its optimization. When progress is made in that area, the methodology can be applied to time-varying examples.

The book is organized as follows. Chapter 1 treats nonlinear control problems in which the cost functional is of the form of a quotient or a product of powers of definite integrals. In Chapter 2, necessary conditions for optimality are developed in the case of linear control problems, with the cost functional being a quotient of definite integrals. Also, an existence theorem is given for the attainment of the optimal cost in a specialized case. Chapter 3 shows the relationship of such cost functionals to the finite-interval H_∞ problem and the application of the results to optimal disturbance rejection. An expression for the variation of the worst-case performance with system parameter variations is derived. The results are developed in the case of time-varying systems. Chapter 4 derives certain necessary conditions that need to be satisfied by the controller which yields maximum disturbance rejection. In Chapter 5, the tools developed in the previous chapters are applied to the finite-interval H_∞ problem, making use of observer-based parametrization of all stabilizing controllers. In Chapter 6, a generalized finite-interval H_∞ problem is treated, and necessary conditions and existence results are given. Also, an expression for the worst-case performance variation is derived in terms of system parameter variations. Finally, Chapter 7 treats the problem of optimal reduction of a high order system to a low order one. The cost functional involved in such reduction is of the form of a quotient of definite integrals. Only Chapters 5 and 7 make use of figures and these are given at the end of the respective chapters.

I take this opportunity to express my gratitude to Mr. Fred Kuster, Head of the Flight Control Branch at the Naval Air Development Center, for his encouragement and support of this research. Thanks are also due to my colleague Marc Steinberg for providing the necessary computational assistance in connection with Chapter 7. Finally, I thank my wife Carol for her love and support.

Warminster, Pennsylvania

M. BALA SUBRAHMANYAM

January, 1990

TABLE OF CONTENTS

Chapter 4

Necessary Conditions for Optimal Disturbance Rejection

in Linear Systems

CHAPTER 7

Model Reduction with a Finite-Interval

H_∞ Criterion

CHAPTER 1

Necessary Conditions for Optimality in
Problems with Nonstandard Cost Functionals

1. INTRODUCTION

Usual formulation of optimal control problems involves minimization of a cost functional which is of the form of a definite integral. In this chapter, we develop necessary conditions for an optimal control in the case of problems in which the cost functional is either a quotient or a product of definite integrals. We call such functionals nonstandard. Preliminary results for problems having a fixed final time and free terminal state are in [1]. Related results can also be found in [2,3]. In this monograph, we consider only fixed final time problems. Problems in which the final time is free are treated in [4]. In Section 5, we discuss the relation of our results to those in [2,3].

In this chapter, results will be derived for nonlinear systems, although the emphasis in susequent chapters is on linear systems. We make use of the Dubovitskii-Milyutin formalism [5,6] to derive the necessary conditions. This formalism is narrated in Section 2. We will not give detailed proofs of the results in Section 2 since a very lucid treatment of the theory is given in [6].

We consider problems where the cost functional is of the form of a quotient in Section 3, and of the form of a product in Section 4. Finally, certain generalizations are considered in Section 5.

2. PRELIMINARIES

Throughout this section, unless otherwise stated, E denotes a linear topological space [7]. Let $F(x)$ be a real-valued function defined on E.

DEFINITION 2.1. A vector h is called *a direction of decrease* of $F(x)$ at a point x_0 if we can find a neighborhood U of h, and two numbers $\alpha(F, x_0, h) < 0$ and $\epsilon_0 > 0$ such that

$$F(x_0 + \epsilon h) \leq F(x_0) + \epsilon \alpha \quad \text{for all} \ \ 0 < \epsilon < \epsilon_0, \ h \in U.$$

DEFINITION 2.2. A subset $K(\bar{x}) \subset E$ is called *a cone with vertex* $\bar{x}$ if $\bar{x} + \rho(x - \bar{x}) \in K(\bar{x}), \rho > 0$, whenever $x \in K(\bar{x})$. A cone $K(\bar{x})$ can always be obtained as a translate $\bar{x} + K(0)$ of a cone $K(0)$ with vertex 0. If in addition, $K(\bar{x})$ is convex, then it is called a *convex cone*.

It is easy to verify that the directions of decrease generate an open cone $\tilde{K}(0)$. The functional $F(x)$ is said to be *regularly decreasing* at x_0 if $\tilde{K}(0)$ is convex.

DEFINITION 2.3. The derivative $F'(x_0, h)$ at a point x_0 in the direction of h is given by

$$F'(x_0, h) = \lim_{\epsilon \to 0+} \frac{F(x_0 + \epsilon h) - F(x_0)}{\epsilon}.$$

The following result can be found in [6].

THEOREM 2.1. *Let E be a Banach Space and $F(x)$ satisfy a local Lipschitz condition at x_0 (i.e., there exist $\epsilon_0 > 0$ and $\beta > 0$ such that $|F(x_1) - F(x_2)| \leq \beta \|x_1 - x_2\|$ for all $\|x_1 - x_0\| \leq \epsilon_0, \|x_2 - x_0\| \leq \epsilon_0$). Assume that $F(x)$ is differentiable at x_0 in any direction, and $F'(x_0, h)$ is convex as a functional of h (i.e., for any $0 \leq \rho \leq 1, F'(x_0, \rho h_1 + (1 - \rho)h_2) \leq \rho F'(x_0, h_1) + (1 - \rho)F'(x_0, h_2))$. Then $F(x)$ is regularly decreasing at x_0, and $\tilde{K}(0) = \{h \mid F'(x_0, h) < 0\}$.*

Proof. Theorem 7.3 of [6].

DEFINITION 2.4. A nonzero continuous linear functional g is said to be a *support functional* for a set A at $x_0 \in A$ if $g(x) \geq g(x_0)$ for all $x \in A$. Under these conditions, the closed hyperplane $H = \{x \mid g(x) = g(x_0)\}$ is called a *supporting hyperplane* for A at the point x_0.

DEFINITION 2.5. Let $Q \subset E$. A vector h is said to be a *feasible direction* for Q at $x_0 \in E$ if there exists a neighborhood U of h such that $x_0 + \epsilon h \in Q$ for all $h \in U$ and all $0 < \epsilon < \epsilon_0$ for some positive number ϵ_0.

It can be easily verified that the feasible directions generate an open cone K_1 with vertex at 0. We say that Q is *regular in feasible directions* at x_0 if $K_1(0)$ happens to be convex.

The dual space of E (the set of all continuous linear functionals on E) is denoted by E^*. The space E^* is a Banach space with the norm $\|g\| = \sup_{\|x\| \leq 1} |g(x)|, g \in E^*$, if E is a normed linear space. If a cone $K(0) \subset E$, the *dual cone* $K^* = \{g \in E^* \mid g(x) \geq 0$ on $K(0)\}$.

We state the following result on dual cones from [6].

THEOREM 2.2. *Let $Q \in E$ be a closed convex set and $x_0 \in Q$. Let Q^* denote the set of all support functionals for Q at x_0 and K_b, the cone of feasible directions for Q at x_0. If $\text{int}(Q) \neq \emptyset$, then $K_b^* = Q^*$.*

Proof. Theorem 10.5 of [6].

DEFINITION 2.6. Let $Q \in E$. A vector h is said to be *a tangent direction* to Q at $x_0 \in E$ if we can find $x(\epsilon) \in Q$ for all ϵ between 0 and some $\epsilon_0 > 0$, such that $x(\epsilon) = x_0 + \epsilon h + r(\epsilon)$. The vector $r(\epsilon)$ is such that for any neighborhood U of 0, $(1/\epsilon)r(\epsilon) \in U$ for all sufficiently small $\epsilon > 0$.

It is easily seen that the tangent directions generate a cone with vertex at 0. We say that Q is *regular in tangent directions* at x_0 if the cone of tangent directions to Q at x_0 is convex.

We now give the fundamental theorem due to Dubovitskii and Milyutin [5,6].

THEOREM 2.3. *Let the functional $F(x)$ assume a local minimum on $Q = \bigcup_{i=1}^{n+1} Q_i$ at a point $x_0 \in Q$. Assume that $F(x)$ is regularly decreasing at x_0, with directions of decrease K_0; the inequality constraints $Q_i, i = 1, \ldots, n$ (to be made precise later) are regular in feasible directions at x_0; the equality constraint Q_{n+1} (to be made precise later) is also regular in tangent directions at x_0. Denote the feasible directions for each $Q_i, i = 1, \ldots, n$, by K_i and the tangent directions for Q_{n+1} at x_0 by K_{n+1}. Then there exist $g_i \in K_i^*, i = 0, 1, \ldots, n + 1$, not all zero, such that*

$$\sum_{i=0}^{n+1} g_i = 0.$$

Proof. Theorem 6.1 of [6].

To add a note on the notation to be used, $C^n(0, T)$ denotes the space of all n-tuples of real-valued continuous functions on $[0, T]$ with sup-norm topology, and $L_\infty^r(0, T)$ represents the space of all r-tuples of essentially bounded real-valued measurable functions on $[0, T]$ with the usual norm topology. If f is a real-valued function with x as one of its arguments, the partial derivative of f with respect to x is represented by f_x. If B is a matrix, B^* denotes its transpose. We already used the superscript $*$ to denote dual spaces, dual cones etc., and hopefully, no confusion arises on account of this dual usage. The symbol $(\cdot, \cdot)$ represents an ordered pair or inner product, whichever is applicable.

For convenience, we state below a result in [6] which will be subsequently used.

LEMMA 2.1. *Let $Q = \{x \in L_\infty^r(0, T) \mid x(t) \in M \text{ for almost all } 0 \le t \le T, M \subset R^r\}, x_0 \in Q$. Then, if the linear functional defined by*

$$g(x) = \int_0^T (a(t), x(t)) \, dt, \quad a \in L_1^r(0, T),$$

is a support to Q at the point x_0, then $(a(t), x(t) - x_0(t)) \ge 0$ for all $x \in M$ and almost all $0 \le t \le T$.

Proof. A simple argument results in the contrapositive. Also, see Example 10.5 of [6].

3. Necessary Conditions for Optimality

In this section, we develop necessary conditions for an optimal control for problems in which the final time is fixed. Consider the system

$$\frac{dx}{dt} = f(x(t), u(t), t) \tag{1}$$

with boundary conditions

$$x(0) = c, \tag{2}$$

$$x(T) = d \quad (T \text{ fixed}), \tag{3}$$

where $x(t) \in R^n, u(t) \in R^r$, and t represent the state vector, the control vector, and time respectively. The problem is to determine the conditions on $x(t) \in C^n(0, T)$ and $u(t) \in L_\infty^r(0, T)$ which minimize

$$F(x, u) = \frac{\int_0^T \phi^1(x(t), u(t), t) \, dt}{\int_0^T \phi^2(x(t), u(t), t) \, dt} \tag{4}$$

(where ϕ^1 and ϕ^2 are scalar functions), under the constraint

$$u(t) \in M \subset R^r \quad \text{for almost all} \quad 0 \le t \le T. \tag{5}$$

The case where the domain of definition of the solution is $[t_0, T], t_0 \neq 0$, can be reduced to the above case by a simple substitution of variables.

Let $f(x, u, t)$ and $\phi^i(x, u, t), i = 1, 2$, be continuous in x and u, measurable in t, and continuously differentiable with respect to x and u. Also let $f_u, f_x, \phi_u^i, i = 1, 2, \phi_x^i, i = 1, 2$, be bounded for all bounded (x, u). The set M is assumed to be convex with $\text{int}(M) \neq \emptyset$.

With the above assumptions, let us state the necessary conditions for an optimal control.

THEOREM 3.1. *Let $x^0(t)$ and $u^0(t)$ be optimal. Assume that $\int_0^T \phi^2(x^0, u^0, t)\,dt > 0$. Also assume that $\int_0^T \phi^1(x^0, u^0, t)\,dt$ and $\int_0^T \phi^2(x^0, u^0, t)\,dt$ exist in the sense of Lebesgue. Then there exist $\psi(t) \in R^n$ and $\lambda_0 > 0, \lambda_0 \in R^1$, not both identically zero, such that*

$$\frac{d\psi}{dt} = -f_x^*(x^0, u^0, t)\psi(t) + \lambda_0\{\phi_x^1(x^0, u^0, t) - \lambda\phi_x^2(x^0, u^0, t)\}, \tag{6}$$

where

$$\lambda = \frac{\int_0^T \phi^1(x^0, u^0, t)\,dt}{\int_0^T \phi^2(x^0, u^0, t)\,dt}, \tag{7}$$

and, moreover,

$$([-f_u^*(x^0, u^0, t)\psi(t) + \lambda_0\{\phi_u^1(x^0, u^0, t) - \lambda\phi_u^2(x^0, u^0, t)\}], u - u^0(t)) \geq 0 \tag{8}$$

for almost all $0 \leq t \leq T$ and all $u \in M$.

Proof. Let $E = C^n(0, T) \times L_\infty^r(0, T)$. Let Q_2 denote the set of all $(x, u) \in E$ satisfying (1), (2), and (3) and Q_1, the set of all pairs satisfying (5). Regarding Q_1 and Q_2 as inequality and equality constraints, respectively, our problem is to minimize (4) on $Q_1 \cap Q_2$.

(a) *Analysis of the functional $F(x, u)$*

By Theorem 2.1, $(\bar{x}(t), \bar{u}(t))$ lies in the cone of directions of decrease if and only if (let $\theta^0(t) = (x^0(t), u^0(t)), \gamma^0(t) = (\theta^0(t), t)$)

$$F'(\theta^0, (\bar{x}, \bar{u})) = \frac{\left[\begin{array}{c}[\int_0^T \phi^2(\gamma^0)\,dt]\int_0^T[(\phi_x^1(\gamma^0), \bar{x}) + (\phi_u^1(\gamma^0), \bar{u})]\,dt \\ -[\int_0^T \phi^1(\gamma^0)\,dt]\int_0^T[(\phi_x^2(\gamma^0), \bar{x}) + (\phi_u^2(\gamma^0), \bar{u})]\,dt\end{array}\right]}{[\int_0^T \phi^2(\gamma^0)\,dt]^2} < 0,$$

provided that the denominator is nonzero. Let $\int_0^T \phi^2(x^0, u^0, t)\,dt \neq 0$. Since the denominator is positive, $(\bar{x}, \bar{u})$ lies in K_0 if and only if (simplifying the notation)

$$\left[\int_0^T \phi^2\,dt\right]\int_0^T[(\phi_x^1, \bar{x}) + (\phi_u^1, \bar{u})]\,dt - \left[\int_0^T \phi^1\,dt\right]\int_0^T[(\phi_x^2, \bar{x}) + (\phi_u^2, \bar{u})]\,dt < 0. \tag{9}$$

Let $\int_0^T \phi^1(x^0, u^0, t)\, dt / \int_0^T \phi^2(x^0, u^0, t)\, dt = \lambda$, and without loss of generality, let

$$\int_0^T \phi^2(x^0, u^0, t)\, dt > 0.$$

Then (9) can be replaced by

$$\int_0^T [(\phi_x^1, \bar{x}) + (\phi_u^1, \bar{u})]\, dt - \lambda \int_0^T [(\phi_x^2, \bar{x}) + (\phi_u^2, \bar{u})]\, dt < 0.$$

By standard arguments (for example, see [6, Theorem 10.2, p. 69]), if $K_0 \neq \emptyset$, then for any $g_0 \in K_0^*$,

$$g_0(\bar{x}, \bar{u}) = -\lambda_0 \Big\{ \int_0^T [(\phi_x^1, \bar{x}) + (\phi_u^1, \bar{u})]\, dt$$

$$- \lambda \int_0^T [(\phi_x^2, \bar{x}) + (\phi_u^2, \bar{u})]\, dt \Big\}, \lambda_0 \geq 0. \tag{10}$$

(b) *Analysis of the constraint Q_1*

The set Q_1 is closed and convex in E since $Q_1 = C^n(0, T) \times Q_1'$, where $Q_1' = \{u(t) \in L_\infty^r(0, T) \mid u(t) \text{ obeys (5)}\}$ is closed and convex in $L_\infty^r(0, T)$ and has nonempty interior. Also, $\text{int}(Q_1) \neq \emptyset$. Let K_1 be the cone of feasible directions for Q_1 at (x^0, u^0). Then if $g_1 \in K_1^*$, it follows that (see Theorem 2.2) $g_1 = (0, g_1')$, where $g_1' \in [L_\infty^r(0, T)]^*$ is a support to Q_1' at u^0.

(c) *Analysis of the constraint Q_2*

Assume that the nondegeneracy condition $f_u^*(x^0, u^0, t)\psi(t) \neq 0$ holds for any nonzero solution $\psi(t)$ of

$$\frac{d\psi}{dt} = -f_x^*(x^0, u^0, t)\psi(t).$$

Then the tangent subspace K_2 at (x^0, u^0) is the set of all pairs such that

$$\frac{d\bar{x}}{dt} = f_x(x^0, u^0, t)\bar{x} + f_u(x^0, u^0, t)\bar{u}, \quad \bar{x}(0) = 0, \tag{11}$$

$$\bar{x}(T) = 0. \tag{12}$$

Let $L_1 \subset E, L_2 \subset E$ denote the sets of all $(\bar{x}, \bar{u})$ satisfying (11) and (12), respectively. Then L_1 and L_2 are subspaces and $K_2 = L_1 \cap L_2$. It is obvious that if $g \in L_2^*$, then $g(\bar{x}, \bar{u}) = (\bar{x}(T), a), a \in R^n$. The space L_2^* is therefore n-dimensional and $L_1^* + L_2^*$ is weak* closed. Here L_1^* and L_2^* are dual cones. It follows that $K_2^* = L_1^* + L_2^*$. Since L_1 is a subspace, for any $g_2 \in L_1^*, g_2(\bar{x}, \bar{u}) = 0$ for all $(\bar{x}, \bar{u}) \in L_1$. As we already know, if $g_3 \in L_2^*$, then $g_3(\bar{x}, \bar{u}) = (\bar{x}(T), a), a \in R^n$.

(d) *Application of Theorem 2.3*

It can be shown that the cone K_1 is convex (see [5,6]). Hence, by Theorem 2.3, there exist $g_0, g_1, g_2, g_3 \in E^*$, not all zero, such that for all $(\bar{x}, \bar{u}) \in E$,

$$g_0 + g_1 + g_2 + g_3 = 0, \tag{13}$$

where g_0 is given by (10), $g_1(\bar{x}, \bar{u}) = g_1'(\bar{u})$ is a support to Q_1' at u^0, $g_2(\bar{x}, \bar{u})$ vanishes for $(\bar{x}, \bar{u})$ satisfying (11) and $g_3(\bar{x}, \bar{u}) = (\bar{x}(T), a), a \in R^n$.

(e) *Analysis of Equation* (13)

Let $\bar{u}$ be arbitrary and $\bar{x}(\bar{u})$ be the corresponding solution of (11). Under these conditions $g_2(\bar{x}, \bar{u}) = 0$, and (13) is equivalent to

$$g_1'(\bar{u}) = \lambda_0 \{ \int_0^T [(\phi_x^1, \bar{x}) + (\phi_u^1, \bar{u})] \, dt$$

$$- \lambda \int_0^T [(\phi_x^2, \bar{x}) + (\phi_u^2, \bar{u})] \, dt \} - (\bar{x}(T), a), \qquad \lambda_0 \geq 0. \tag{14}$$

Let $\psi(t)$ be the solution of (6) with the boundary condition $\psi(T) = a$. Then it follows that

$$\lambda_0 \int_0^T [(\phi_x^1, \bar{x}) - \lambda(\phi_x^2, \bar{x})] \, dt - (\bar{x}(T), a) = - \int_0^T (f_u^*(x^0, u^0, t)\psi, \bar{u}) \, dt.$$

Hence

$$g_1'(\bar{u}) = \int_0^T ([-f_u^*(x^0, u^0, t)\psi + \lambda_0\{\phi_u^1(x^0, u^0, t) - \lambda\phi_u^2(x^0, u^0, t)\}], \bar{u})\, dt,$$

where $\bar{u}$ is arbitrary and $g_1'(\bar{u})$ is a support to Q_1' at u^0. Now, using Lemma 2.1, we have

$$([-f_u^*(x^0, u^0, t)\psi(t) + \lambda_0\{\phi_u^1(x^0, u^0, t) - \lambda\phi_u^2(x^0, u^0, t)\}], u - u^0(t)) \geq 0$$

for almost all $0 \leq t \leq T$, and all $u \in M$, i.e., (8) is satisfied.

If $\lambda_0 = 0$ and $\psi(t) \equiv 0$, then we would have $g_i = 0, i = 0, 1, 2, 3$, and this contradicts Theorem 2.3.

(f) *Analysis of Exceptional Cases*

We show that even if $K_0 = \emptyset$ and system (11) is degenerate, the conclusions of Theorem 3.1 are valid. If $K_0 = \emptyset$, then

$$\int_0^T [(\phi_x^1, \bar{x}) + (\phi_u^1, \bar{u})]\, dt - \lambda \int_0^T [(\phi_x^2, \bar{x}) + (\phi_u^2, \bar{u})]\, dt = 0.$$

Choose $\lambda_0 = 1$ and $\psi(T) = 0$. Then

$$-f_u^*(x^0, u^0, t)\psi(t) + \lambda_0\{\phi_u^1(x^0, u^0, t) - \lambda\phi_u^2(x^0, u^0, t)\} = 0$$

for almost all $0 \leq t \leq T$, and hence, (8) is satisfied. If (11) is degenerate, choosing $\lambda_0 = 0$ we get a nonzero solution $\psi(t)$ of (6) with $-f_u^*(x^0, u^0, t)\psi(t) \equiv 0$.   □

Thus the proof of Theorem 3.1 is complete. If the boundary conditions are such that $x(0) \in S_1, x(T) \in S_2$, where S_1 and S_2 are smooth manifolds in R^n, then the results of Theorem 3.1 are still valid, with the added transversality conditions: $\psi(0)$ and $\psi(T)$ must be orthogonal to the tangent subspaces of S_1 at $x^0(0)$ and S_2 at $x^0(T)$, respectively.

4. COST FUNCTIONAL OF THE FORM OF A PRODUCT

Let us now consider the following problem: Find $(x(t), u(t)) \in C^n(0,T) \times L^r_\infty(0,T)$ that minimizes

$$F(x,u) = \left(\int_0^T \phi^1(x(t),u(t),t)\,dt \right) \left(\int_0^T \phi^2(x(t),u(t),t)\,dt \right) \tag{15}$$

under the constraints (1), (2), (3), and (5).

The assumptions for this section are the same as those in Section 3. Following similar procedure as in Section 3, the following necessary conditions can be derived.

THEOREM 4.1. *Let* $x^0(t)$, *and* $u^0(t)$ *be optimal. Assume that* $\int_0^T \phi^2(x^0,u^0,t)\,dt > 0$. *Also let* $\int_0^T \phi^1(x^0,u^0,t)\,dt$ *and* $\int_0^T \phi^2(x^0,u^0,t)\,dt$ *be finite. Then there exist* $\psi(t), \lambda_0 \geq 0$, *not both identically zero, such that*

$$\frac{d\psi}{dt} = -f_x^*(x^0,u^0,t)\psi(t) + \lambda_0\{\phi_x^1(x^0,u^0,t) + \lambda\phi_x^2(x^0,u^0,t)\},$$

and

$$([-f_u^*(x^0,u^0,t)\psi(t) + \lambda_0\{\phi_u^1(x^0,u^0,t) + \lambda\phi_u^2(x^0,u^0,t)\}], u - u^0(t)) \geq 0$$

for all $u \in M$ *and almost all* $0 \leq t \leq T$,

where

$$\lambda = \frac{\int_0^T \phi^1(x^0,u^0,t)\,dt}{\int_0^T \phi^2(x^0,u^0,t)\,dt}.$$

5. CERTAIN GENERALIZATIONS

Work similar to ours employing variational techniques can be found in [2,3]. The problem treated there involves fixed initial and final times and states. We showed in [4] that similar results can be obtained in the case where the final time is not fixed. Also our results are applicable to the case involving control constraints and in general, we assume less smoothness on the functions $f(x, u, t)$ and $\phi^i(x, u, t), i = 1, 2$.

In [2,3], Miele considers an optimal problem involving products of powers of a finite number of functionals. We will extend our results to this case in the present section. The problem will be the same as the one considered in Section 3 with the cost functional replaced by

$$F(x, u) = \left(\int_0^T \phi^1(x(t), u(t), t) \, dt \right)^{\alpha_1} \left(\int_0^T \phi^2(x(t), u(t), t) \, dt \right)^{\alpha_2}, \tag{19}$$

where $\alpha_1, \alpha_2 \in R^1$. We impose the same conditions on f, ϕ^1, and ϕ^2 as those in Section 3.

Let $(x^0(t), u^0(t))$ be a solution to the above problem. Assume further that $0 < \int_0^T \phi^i(x^0, u^0, t) \, dt < \infty, i = 1, 2$. Then $(x^0(t), u^0(t))$ solves the equivalent problem with the alternate cost functional

$$G(x, u) = \ln F(x, u) = \sum_{i=1}^{2} \alpha_i \ln \left(\int_0^T \phi^i(x(t), u(t), t) \, dt \right). \tag{20}$$

Note that

$$G'((x^0, u^0), (\bar{x}, \bar{u})) = \sum_{i=1}^{2} \frac{\alpha_i \int_0^T [(\phi^i_x, \bar{x}) + (\phi^i_u, \bar{u})] \, dt}{\int_0^T \phi^i(x^0, u^0, t) \, dt},$$

and $G(x, u)$ satisfies the local Lipschitz condition mentioned in Theorem 2.1. Now, mimicking the procedure in Section 3, we get the following lemma.

LEMMA 5.1. *If $(x^0(t), u^0(t))$ is optimal, then there exist $\psi(t), \lambda_0 \geq 0$, not both identically zero, such that*

$$\frac{d\psi}{dt} = -f_x^*(x^0, u^0, t)\psi + \lambda_0 \sum_{i=1}^{2} \alpha_i \frac{\phi^i_x(x^0, u^0, t)}{\int_0^T \phi^i(x^0, u^0, t) \, dt},$$

and

$$([-f_u^*(x^0, u^0, t)\psi(t) + \lambda_0 \{\sum_{i=1}^{2} \frac{\alpha_i \phi^i_u(x^0, u^0, t)}{\int_0^T \phi^i(x^0, u^0, t) \, dt}\}], u - u^0(t)) \geq 0$$

for almost all $0 \leq t \leq T$ and all $u \in M$.

It can be seen that similar results can be obtained if (19) involves a product of powers of more than two (but a finite number) of definite integrals.

We make use of the above results in later chapters to solve certain optimal control problems. For now, we present a simple scalar example. We wish to find $x(t)$ which yields the minimum of $(\int_0^1 (\dot{x})^2 \, dt)(\int_0^1 x^2 \, dt)^{-1}$ under the boundary conditions $x(0) = x(1) = 0$. Letting $\dot{x} = u(t)$, the functional to be minimized becomes $(\int_0^1 u^2 \, dt)(\int_0^T x^2 \, dt)^{-1}$. Applying Theorem 3.1, the solutions that satisfy the boundary conditions are

$$x^0(\lambda, t) = A \sin(\lambda^{1/2} t), \quad A \neq 0,$$

where $\lambda = n^2 \pi^2, n = 1, 2, \ldots$. The cost for these curves is given by

$$F(x^0, u^0) = \frac{\int_0^1 A^2 n^2 \pi^2 \cos^2(n\pi t) \, dt}{\int_0^1 A^2 \sin^2(n\pi t) \, dt} = n^2 \pi^2.$$

The least possible value of F is achieved for $n = 1$ when $x(t) = A \sin \pi t, A \neq 0$.

REFERENCES

[1] M. B. SUBRAHMANYAM AND E. D. EYMAN, "Optimization with nonstandard cost functionals," Proc. 13th Annual Allerton Conf., University of Illinois, 1975.

[2] A. MIELE, The extremization of products of powers of functionals and its application to aerodynamics, *Astronaut. Acta* **12**, No. 1, 1967, pp. 47-51.

[3] ――――――, On the minimization of the product of the powers of several integrals, *J. Optimization Theory Appl.* **1**, No. 2, 1967, pp. 70-82.

[4] M. B. SUBRAHMANYAM, Necessary conditions for minimum in problems with non-standard cost functionals, *J. Math. Anal. Appl.* **60**, No. 3, 1977, pp. 601-616.

[5] A. YA. DUBOVITSKII AND A. A. MILYUTIN, Extremum problems in the presence of restrictions [English translation], *U.S.S.R. Comput. Math. Math. Phys.* 5, No. 3, 1965, pp. 1-80.

[6] I. V. GIRSANOV, "Lecture Notes in Economics and Mathematical Systems," No. 67, Springer-Verlag, New York, 1972.

[7] G. KÖTHE, "Topological Vector Spaces," Vol. I, Springer-Verlag, New York, 1969.

CHAPTER 2

Linear Control Problems and an Existence Theorem

1. INTRODUCTION

In Chapter 1, we derived necessary conditions for an optimal control in the case of nonlinear problems with a nonstandard cost functional. In this chapter we derive an existence result for time-varying linear systems. The existence theorem will be proved in a general case in which the interval of interest need not be finite. We also derive necessary conditions for an optimal control. Even though the results of Chapter 1 can be used to derive the necessary conditions, we derive these in an independent fashion, since certain details will be better amplified in the proof given here. These results will be used in subsequent chapters to address the issues of disturbance rejection and performance robustness.

The material in this chapter follows closely certain sections of [1]. Related material can also be found in [2]. The proof of Theorem 2.1 utilizes the concept of weak convergence of a certain sequence, and for a definition of this term, see [3].

2. AN EXISTENCE THEOREM

To be more specific, consider the n-dimensional system

$$\dot{x} = A(t)x + B(t)u, \quad x(t_0) = 0, \tag{1}$$

where $t \in [t_0, T], T \leq \infty$. We impose a finite number of constraints on the trajectory x and the control u, such as, $\lim_{t \to T} x(t) = 0, \int_{t_0}^{T} u^p \, dt = 0, p > 0$, and so on. We lay some restrictions on these constraints later.

The functional to be minimized is

$$F(x, u) = \frac{\int_{t_0}^T \phi^1(u, t)\, dt}{[\int_{t_0}^T \phi^2(x, t) f(t)\, dt]^\alpha},$$ (2)

where $\alpha > 0$, $f(t) \geq 0$ is measurable and u is a measurable control. We make the following assumptions.

(a) $A(t)$ and $B(t)$ are continuous $n \times n$ and $n \times r$ matrix functions respectively.

(b) For $i = 1, 2$, ϕ^i is continuous in x, u and t. Also, for each t, ϕ^1 is convex in u.

(c) Admissible controls are measurable functions on $[t_0, T]$ such that $\int_{t_0}^T \phi^1\, dt < \infty$.

(d) $\phi^1(u, t) \geq a|u|^p$, $a > 0, p > 1$, and $\phi^2(x(t), t) \geq 0$ along any $x(t)$ which is the response to some admissible $u(t)$.

(e) For each $K < \infty$, there is an integrable $g_K(t)$ such that, if $\|u\|_p \leq K$, then

$$|\phi^2(x(t), t) f(t)| \leq g_K(t)$$ (3)

almost everywhere on $[t_0, T]$ for any admissible u whose trajectory obeys any constraints imposed.

(f) There exists $k > 0$ such that, for every $c \geq 0$,

$$\phi^1(cu, t) = c^k \phi^1(u, t),$$

$$\phi^2(cx, t) = c^{k/\alpha} \phi^2(x, t).$$ (4)

By (3), this assumption implies that for every $c > 0, F(cx, cu) = F(x, u)$.

(g) There exists an admissible control the trajectory of which satisfies the imposed constraints and is such that

$$0 < \int_{t_0}^T \phi^2(x, t) f(t)\, dt < \infty.$$

We call a constraint *regular* if the following two conditions hold:

(1) (x, u) satisfies the constraint $\Rightarrow$ (cx, cu) satisfies the constraint for every $c > 0$.

(2) Let $(x^1, u^1), (x^2, u^2), \cdots$ be admissible pairs such that $u^i \to u^0$ weakly in $L_p(t_0, T) = \{u = (u_1, \ldots, u_r) : [t_0, T] \to R^r \mid \|u\|^p = \int_{t_0}^T |u|^p \, dt < \infty\}$. Suppose (x^n, u^n) satisfies the constraint for each $n \geq 1$. Then (x^0, u^0) obeys the constraint. (It is shown in the proof of Theorem 2.1 that u^0 is necessarily admissible.)

PROPOSITION 2.1. *Consider all pairs (x, u) that obey (1) and the constraints. Assume that all the constraints are regular, and let*

$$\lambda = \inf_{(x,u)} F(x, u) = \inf_{(x,u)} \frac{\int_{t_0}^T \phi^1(u, t) \, dt}{\left[\int_{t_0}^T \phi^2(x, t) f(t) \, dt\right]^\alpha}. \tag{5}$$

(λ is well defined by assumptions (c) and (g).) Also, let

$$\inf_u \int_{t_0}^T \phi^1(u, t) \, dt = J \quad \text{subject to} \quad \left[\int_{t_0}^T \phi^2(x, t) f(t) \, dt\right]^\alpha = M > 0. \tag{6}$$

Then $\lambda = J/M$.

Proof. Clearly $J/M \geq \lambda$. To reverse the inequality, let $\tilde{u}$ be such that $F(\tilde{x}, \tilde{u}) \leq \lambda + \epsilon$ for some $\epsilon \geq 0$. Let $\left[\int_{t_0}^T \phi^2(\tilde{x}, t) f(t) \, dt\right]^\alpha = \tilde{M}$ ($< \infty$ by assumptions (c), (d), and (e)), and $\mu = (M/\tilde{M})^{1/k}$. Then $(\mu\tilde{x}, \mu\tilde{u})$ obeys all the constraints by the regularity of the constraints, and by assumption (f), $\left[\int_{t_0}^T \phi^2(\mu\tilde{x}, t) f(t) \, dt\right]^\alpha = M$ and $F(\mu\tilde{x}, \mu\tilde{u}) \leq \lambda + \epsilon$. By (6), $J/M \leq \lambda + \epsilon$. Since ϵ is arbitrary, the conclusion of the proposition follows. $\square$

THEOREM 2.1. *Consider the system (1) and (2) along with assumptions (a)-(g). Also assume that the constraints on x and u are regular. Then there exists a control among all admissible controls that minimizes (2).*

Proof. By Proposition 2.1, it is sufficient to exhibit a minimizing control among all admissible controls for which $\left[\int_{t_0}^T \phi^2(x, t) f(t) \, dt\right]^\alpha = M > 0$ and the trajectories of which satisfy (1) and all the constraints. Let $J = \inf_u \int_{t_0}^T \phi^1(u, t) \, dt$ subject to $\left[\int_{t_0}^T \phi^2(x, t) f(t) \, dt\right]^\alpha =$

M. Choose $\{(x^i, u^i)\}$ such that $\lim_{i \to \infty} \int_{t_0}^T \phi^1(u^i, t) \, dt = J$ with $\left[\int_{t_0}^T \phi^2(x^i, t) f(t) \, dt \right]^\alpha = M$

for each i. By assumption (d), $\{u^i\}$ form a bounded sequence in $L_p(t_0, T)$, and hence a

subsequence, still denoted by $\{u^i\}$, converges weakly to some u^0 in $L_p(t_0, T)$. Let x^0 be

the response of (1) to u^0. By assumption (a) and by the weak convergence, $x^i(t) \to x^0(t)$

for all $t \in [t_0, T)$. By the regularity of the constraints, $x^0(t)$ obeys all the constraints.

Assumption (b) $\Rightarrow \phi^2(x^i(t), t)$ converges to $\phi^2(x^0(t), t)$ for all $t \in [t_0, T)$. Since $\|u^i\|_p \leq K$

for some $K < \infty$, by assumption (e) and by the Lebesgue dominated convergence theorem,

$$\left[\int_{t_0}^T \phi^2(x^0(t), t) f(t) \, dt \right]^\alpha = \lim_{i \to \infty} \left[\int_{t_0}^T \phi^2(x^i(t), t) f(t) \, dt \right]^\alpha = M.$$

If $T < \infty$, then we have by assumption (b) (see [4,p.209]),

$$\int_{t_0}^T \phi^1(u^0, t) \, dt \leq \liminf_{i \to \infty} \int_{t_0}^T \phi^1(u^i, t) \, dt = J.$$

If $T = \infty$, $[t_0, T) = \bigcup_{j=0}^\infty [t_0 + j, t_0 + j + 1]$. We have on each subinterval,

$$a_j = \int_{t_0+j}^{t_0+j+1} \phi^1(u^0, t) \, dt \leq \liminf_{i \to \infty} \int_{t_0+j}^{t_0+j+1} \phi^1(u^i, t) \, dt = a_{ji}.$$

Hence by Fatou's lemma

$$\sum_{j=0}^\infty a_j \leq \sum_{j=0}^\infty \liminf_{i \to \infty} a_{ji} \leq \liminf_{i \to \infty} \sum_{j=0}^\infty a_{ji} = J.$$

Thus the proof is complete.   □

3. NECESSARY CONDITIONS FOR OPTIMALITY

In this section, we develop necessary conditions for an optimal control in the finite

interval case with integrable $f(t)$ (see (2)). Let $I = [t_0, T]$, $T < \infty$, and $C(I)$ be the space of all continuous $x(t) = (x_1(t), \ldots, x_n(t)) : I \to R^n$ such that $\|x\| = \max\{\|x_1\|, \cdots, \|x_n\|\}$, where $\|x_i\| = \sup_{[t_0, T]} |x_i(t)|$. Once again, consider the n-dimensional system

$$\frac{dx}{dt} = A(t)x + B(t)u, \tag{7}$$

with

$$x(t_0) = c, \qquad x(T) = d, \qquad T < \infty, \tag{8}$$

$$F(x, u) = \frac{\int_{t_0}^{T} \phi^1(u, t)\, dt}{[\int_{t_0}^{T} \phi^2(x, t)f(t)\, dt]^\alpha}, \tag{9}$$

where $f(t)$ is integrable and $\alpha \in R$. The result in the case where ϕ^2 is also a function of u will be stated as a corollary at the end of this section.

We make the following assumptions.

(a) $A(t)$ and $B(t)$ are continuous $n \times n$ and $n \times r$ matrix functions respectively.

(b) Admissible controls are measurable functions such that $\int_{t_0}^{T} \phi^1(u, t)\, dt < \infty$.

(c) $\phi^1(u, t) \geq a|u|^p, a > 0, p > 1$, and $\phi^2(x(t), t) \geq 0$ a.e. (almost everywhere) on $[t_0, T]$ for any trajectory $x(t)$ which is the response to some $u \in L_p(t_0, T)$.

(d) Let ϕ^1 be continuously differentiable in u and ϕ^2 be continuously differentiable in x, and both be measurable in t. Moreover, let $\phi^1(u(t), t) \in L^q(t_0, T)$ for all $u \in L_p(t_0, T), (1/p) + (1/q) = 1$. Also, let ϕ_x^2 be bounded for bounded x, the bound being uniform for almost all t.

(e) Let ϕ_u^1 and ϕ_x^2 be locally Lipschitzian in u and x respectively, i.e., there exist $\delta > 0, K_1, K_2 > 0$ depending on (x, u) such that for all $\bar{x} \in C(I)$ and $\bar{u} \in L_p(t_0, T)$ with $\|\bar{x}\| \leq \delta, \|\bar{u}\|_p \leq \delta$, we have

$$|\phi_x^2(x + \bar{x}, t) - \phi_x^2(x, t)| \leq K_1\|\bar{x}\|$$

almost everywhere, and

$$\|\phi_u^1(u + \bar{u}, t) - \phi_u^1(u, t)\|_q \leq K_2 \|\bar{u}\|_p.$$

(f) $(x^0(t), u^0(t))$ minimizes (9) subject to (7) and (8)

$$\Rightarrow 0 < \int_{t_0}^T \phi^1(u^0(t), t) \, dt < \infty, \quad 0 < \int_{t_0}^T \phi^2(x^0(t), t) f(t) \, dt < \infty.$$

(g) The pair $(A(t), B(t))$ is completely controllable.

By assumption (f), we can consider the alternate cost functional

$$G(x, u) = \ln \int_{t_0}^T \phi^1 \, dt - \alpha \ln \int_{t_0}^T \phi^2 f(t) \, dt \tag{10}$$

in place of (9). In order to establish our necessary conditions, we will make use of the Dubovitskii-Milyutin theorem [Ch. 1, Thm. 2.3]. For the definitions of various terms, we refer the reader to Chapter 1. If K is a cone in a Banach space E, we mean by the dual cone K^* the set $\{g \in E^* \mid g(x) \geq 0 \text{ for all } x \in K\}$. Note that the superscript $*$ also denotes a matrix transpose. Now we state the necessary conditions for an optimal control.

THEOREM 3.1. *Consider the system (7)-(9) along with assumptions (a)-(g). Suppose that $(x^0(t), u^0(t))$ minimizes (9). Then there exists $\psi(t) \in C(I)$ such that*

$$\frac{d\psi}{dt} = -A^*(t)\psi - \lambda \phi_x^2(x^0, t) f(t), \tag{11}$$

where

$$\lambda = \frac{\int_{t_0}^T \phi^1(u^0, t) \, dt}{\int_{t_0}^T \phi^2(x^0, t) f(t) \, dt} \tag{12}$$

and

$$\phi_u^1(u^0(t), t) - \alpha B^*(t)\psi(t) = 0 \quad \text{a.e. on } [t_0, T]. \tag{13}$$

Proof. Let $E = C(I) \times L_p(I)$. Our admissible controls form a subset of $L_p(I)$. But by assumption (f), the optimal cost is finite and hence we can regard u^0 to be optimal with respect to those controls in $L_p(I)$ whose trajectories obey (7) and (8). Thus, we take our space of controls to be $L_p(I)$.

(a) *Cone of directions of decrease.* By assumptions (d) and (e), the Fréchet derivative of the functional G in (10) is given by

$$G'(x^0, u^0)(x, u) = \frac{\int_{t_0}^T (\phi_u^1(u^0, t), u)\, dt}{\int_{t_0}^T \phi^1(u^0, t)\, dt} - \alpha \frac{\int_{t_0}^T (\phi_x^2(x^0, t), x) f(t)\, dt}{\int_{t_0}^T \phi^2(x^0, t) f(t)\, dt}. \tag{14}$$

By [5, Thm. 7.5], $(x(t), u(t))$ lies in the cone K_0 of directions of decrease in E if and only if $G'(x^0, u^0)(x, u) < 0$. By assumption (f), $(x, u) \in K_0$ if and only if

$$\int_{t_0}^T (\phi_u^1(u^0, t), u)\, dt - \alpha\lambda \int_{t_0}^T (\phi_x^2(x^0, t), x) f(t)\, dt < 0, \tag{15}$$

where λ is defined by (12). If $K_0 \neq \emptyset$, then by [5, Thm. 10.2], for any $g_0 \in K_0^*$,

$$g_0(x, u) = -\lambda_0 \left\{ \int_{t_0}^T (\phi_u^1(u^0, t), u)\, dt - \alpha\lambda \int_{t_0}^T (\phi_x^2(x^0, t), x) f(t)\, dt \right\}, \qquad \lambda_0 \geq 0. \tag{16}$$

(b) *Cone of tangent directions.* To find the tangent directions in E at (x^0, u^0), we will apply the results of [5, Lecture 9]. Let

$$Q = \{(x, u) \in E \mid x(t) = \Phi(t)c + \Phi(t) \int_{t_0}^t \Phi^{-1}(s)B(s)u(s)\, ds, t_0 \leq t \leq T, x(T) = d\}, \tag{17}$$

where $\Phi(t)$ is a fundamental matrix of $\dot{y} = A(t)y$ with $\Phi(t_0) = I$, and let

$$P(x, u) = \left(x(t) - \Phi(t)c - \Phi(t) \int_{t_0}^t \Phi^{-1}(s)B(s)u(s)\, ds, x(T) \right), \tag{18}$$

which maps E into $C(I) \times R^n$. Also

$$P'(x^0, u^0)(x, u) = \left(x(t) - \Phi(t) \int_{t_0}^t \Phi^{-1}(s)B(s)u(s)\, ds, x(T) \right), \tag{19}$$

where $P'(x^0, u^0) : E \to C(I) \times R^n$. We wish to show that $P'(x^0, u^0)$ is onto.

Let $(a(t), b) \in C(I) \times R^n$. Since $(A(t), B(t))$ is completely controllable, select $\tilde{u} \in L_p(I)$ such that

$$\Phi(t) \int_{t_0}^{t} \Phi^{-1}(s) B(s) \tilde{u}(s) \, ds = b - a(T).$$

Set $\tilde{x}(t) = \Phi(t) \int_{t_0}^{t} \Phi^{-1}(s) B(s) \tilde{u}(s) \, ds + a(t)$. Then $P'(x^0, u^0)(\tilde{x}, \tilde{u}) = (a(t), b)$. By [5, Thm. 9.1] the set K_1 of tangent directions at (x^0, u^0) is given by $\{(x, u) \in E \mid P'(x^0, u^0)(x, u) = 0\}$. Thus K_1 consists of all (x, u) satisfying

$$\frac{dx}{dt} = A(t)x + B(t)u, \qquad x(t_0) = 0, \tag{20}$$

$$x(T) = 0. \tag{21}$$

Let $L_1 \subset E$ denote pairs satisfying (20) and $L_2 \subset E$ the set of (x, u) satisfying (21). It follows that (see [5,Lecture 12]) $K_1^* = L_1^* + L_2^*$, and if $g_2 \in L_2^*$, then $g_2(x, u) = a^* x(T)$ for some $a \in R^n$. If $g_1 \in L_1^*$, then $g_1(x, u) = 0$ for all $(x, u) \in L_1$, since L_1 is a subspace.

(c) *Application of Dubovitskii-Milyutin theorem.* The theorem [Ch. 1, Thm. 2.3] states that there exist $g_0 \in K_0^*, g_1 \in L_1^*$, and $g_2 \in L_2^*$, not all zero, such that for all $(x, u) \in E$,

$$g_0(x, u) + g_1(x, u) + g_2(x, u) = 0. \tag{22}$$

Let u be arbitrary and x be a solution of (20) for this u. Then $g_1(x, u) = 0$, and hence

$$-\lambda_0 \Big\{ \int_{t_0}^{T} (\phi_u^1, u) \, dt - \alpha\lambda \int_{t_0}^{T} (\phi_x^2, x) f(t) \, dt \Big\} + a^* x(T) = 0, \qquad \lambda_0 \geq 0. \tag{23}$$

The scalar λ_0 has to be positive, because if $\lambda_0 = 0$, then from (23), $a^* x(T) = 0$. If $a = 0$, we would have $g_1 = g_2 = g_3 = 0$ by (22), which is not possible. If $a \neq 0$, by the complete controllability of the pair $(A(t), B(t))$ in (20), we can select some $x(t)$ for which $x(T) = a$, which gives the contradiction that $a^*a = 0$. Hence $\lambda_0 > 0$ and (23) becomes

$$\int_{t_0}^{T} (\phi_u^1, u) \, dt - \alpha\lambda \int_{t_0}^{T} (\phi_x^2, x) f(t) \, dt - \frac{1}{\lambda_0} a^* x(T) = 0. \tag{24}$$

Define ψ by

$$\frac{d\psi}{dt} = -A^*\psi - \lambda\phi_x^2(x^0, t)f(t), \quad \psi(T) = \frac{a}{\alpha\lambda_0}. \tag{25}$$

Then

$$\lambda \int_{t_0}^{T} (\phi_x^2 f(t), x) \, dt = -\int_{t_0}^{T} \left(\frac{d\psi}{dt} + A^*\psi, x\right) dt$$

$$= -\frac{a^*}{\alpha\lambda_0} x(T) + \int_{t_0}^{T} (\psi, B(t)u) \, dt. \tag{26}$$

Thus (24) becomes

$$\int_{t_0}^{T} (\phi_u^1 - \alpha B^*\psi, u) \, dt = 0 \tag{27}$$

for arbitrary u. Hence

$$\phi_u^1 - \alpha B^*\psi = 0 \quad \text{a.e. on } [t_0, T]. \tag{28}$$

(d) *Case when $K_0 = \emptyset$.* If $K_0 = \emptyset$, then

$$\int_{t_0}^{T} (\phi_u^1, u) \, dt - \alpha\lambda \int_{t_0}^{T} (\phi_x^2, x) f(t) \, dt = 0 \tag{29}$$

for all $(x, u) \in E$, and we can proceed as above, letting $\psi(t) = 0$. $\square$

It is possible to extend our necessary conditions to more general functionals than in (9); for example, as in Corollary 3.1, to functionals of the form

$$F(x, u) = \frac{\int_{t_0}^{T} \phi^1(u, t) \, dt}{\left[\int_{t_0}^{T} \phi^2(x, u, t) f(t) \, dt\right]^\alpha}. \tag{30}$$

COROLLARY 3.1. *Consider the system (1)-(2) along with the cost functional given by (30). Modify assumptions (a)-(g) in the following manner.*

(i) *In assumption* (d) *assume that* ϕ_x^2 *is continuously differentiable in* (x, u) *and measurable in* t. *Also,* $\phi_u^2 \in L_q(t_0, T), (1/p) + (1/q) = 1$.

(ii) *Replace the first inequality in assumption* (e) *by*

$$\text{ess sup}_{[t_0, T]}|\phi_x^2(x + \bar{x}, u + \bar{u}, t) - \phi_x^2(x, u, t)| \leq K_1(\|\bar{x}\| + \|\bar{u}\|_p),$$

where the symbol ess sup [3] *denotes the essential supremum of the function. If* (x^0, u^0) *minimizes* (30), *then there exists* $\psi(t) \in C(I)$ *such that*

$$\frac{d\psi}{dt} = -A^*(t)\psi - \lambda\phi_x^2(x^0, u^0, t)f(t), \tag{31}$$

where

$$\lambda = \frac{\int_{t_0}^T \phi^1(u^0, t) \, dt}{\int_{t_0}^T \phi^2(x^0, u^0, t)f(t) \, dt} \tag{32}$$

and

$$\phi_u^1(u^0(t), t) - \alpha\lambda\phi_u^2(x^0(t), u^0(t), t) - \alpha B^*(t)\psi(t) = 0 \ a.e. \ on \ [t_0, T]. \tag{33}$$

It is possible to further extend the results of this chapter to systems described by a linear operator equation [6].

REFERENCES

[1] M. B. SUBRAHMANYAM, On applications of control theory to integral inequalities: II, *SIAM J. Control Optim.* **19**, No. 4, 1981, pp. 479-489.

[2] ——————— , On applications of control theory to integral inequalities, *J. Math. Anal. Appl.* **77**, No. 1, 1980, pp. 47-59.

[3] N. DUNFORD AND J. T. SCHWARTZ, "Linear Operators, Part 1," Interscience, New York, 1959.

[4] E. B. LEE AND L. MARKUS, "Foundations of Optimal Control Theory," John Wiley, New York, 1967.

[5] I. V. GIRSANOV, "Lecture Notes in Economics and Mathematical systems," No. 67, Springer-Verlag, New York, 1972.

[6] M. B. SUBRAHMANYAM, On integral inequalities associated with a linear operator equation, *Proc. Amer. Math. Soc.* **92**, No. 3, 1984, pp. 342-346.

CHAPTER 3

Optimal Disturbance Rejection and Performance Robustness
in Linear Systems

ABSTRACT

In this chapter a method is proposed for the optimal design of regulators and ob-
servers from the disturbance rejection and robust performance points of view. For a given
set of system parameters, we obtain a measure of the disturbance rejection capacity of
the system or observer. Optimization routines need to be employed to select control or
observer gains which maximize the disturbance rejection capacity. The general case of
time-varying linear systems is considered and time-domain techniques are employed. Also
the problem of achieving maximum performance as well as required robustness in the presence of parameter uncertainties is considered. An expression is derived for the variation
of performance with parameter changes. The methodology has connections to the H_∞
methods in the case of time-invariant systems. An application to an aircraft wing leveler
system is given to illustrate the methodology.

1. INTRODUCTION

It is important that in the case of a disturbance, the system error due to the disturbance be small. Disturbance rejection is an important factor in the design of flight control
systems for gust load alleviation. Also in several situations, such as an automatic landing
and low altitude high speed terrain following, the effect of disturbances on the response of
the airplane needs to be small. In this chapter we give a measure of the disturbance rejection capacity of a system or an observer by solving an optimal control problem. Control or

observer gains can be selected using an optimization routine to maximize the disturbance rejection capacity. The mathematical theory behind the method is given in Refs. 1 and 2, and is presented in Chapters 1 and 2 of this monograph. Additional related material can be found in Refs. 3-6. The problem can be reformulated in terms of the H_∞ control theory [7,8] for time-invariant systems and more details on this will be given in Section 5.

We now present three problems for which the method is applicable.

PROBLEM 1. (REGULATOR)

Consider the system

$$\dot{x} = F(t)x + B(t)u + G(t)v, \qquad x(t_0) = 0, \tag{1}$$

$$u = C(t)x, \tag{2}$$

where $v(t)$ is a disturbance and $C(t)$ stabilizes (1).

Assign to the above equations the performance index given by

$$J(C, v) = \frac{\int_{t_0}^T v^*(t)R(t)v(t)\, dt}{\int_{t_0}^T \{x^*(t)Q(t)x(t) + u^*(t)U(t)u(t)\}\, dt}, \tag{3}$$

which needs to be maximized by choosing $v(t)$. Throughout this chapter, the superscript $*$ denotes a matrix or vector transpose. Let $\lambda = \inf_v J(C, v)$. Since the state is to be regulated, λ gives a measure of the disturbance rejection capacity of the system for a particular $C(t)$ in (2). Now choose $C(t)$ to maximize λ.

PROBLEM 2. (OBSERVER)

Consider

$$\dot{x} = F(t)x + G(t)v, \qquad x(t_0) = 0, \tag{4}$$

$$z(t) = H(t)x(t), \tag{5}$$

where $v(t)$ and $z(t)$ represent the disturbance input and the output vector respectively. The observer is given by

$$\dot{\hat{x}} = F(t)\hat{x} + L(t)(z(t) - H(t)\hat{x}(t)). \tag{6}$$

Let $e(t) = x(t) - \hat{x}(t)$. Assuming $e(t_0) = 0$, we have

$$\dot{e} = (F(t) - L(t)H(t))e + G(t)v, \qquad e(t_0) = 0. \tag{7}$$

Assume that $L(t)$ stabilizes (7). Let the disturbance rejection capacity λ of the observer be defined by

$$\lambda = \inf_v \frac{\int_{t_0}^T v^*(t)R(t)v(t)\,dt}{\int_{t_0}^T \{e^*(t)Q(t)e(t) + e^*(t)H^*(t)L^*(t)\tilde{R}(t)L(t)H(t)e(t)\}\,dt}. \tag{8}$$

Now choose the observer gain $L(t)$ to maximize λ.

PROBLEM 3. (REGULATOR WITH RECONSTRUCTED STATE)

After $L(t)$ is chosen as in Problem 2, consider the system

$$\dot{x} = F(t)x + B(t)u + G(t)v, \qquad x(t_0) = 0, \tag{9}$$

$$\dot{\hat{x}} = F(t)\hat{x} + B(t)u + L(t)(z(t) - H(t)\hat{x}(t)), \qquad \hat{x}(t_0) = 0. \tag{10}$$

Let

$$u(t) = C(t)\hat{x}(t) \tag{11}$$

and

$$\lambda = \inf_v \frac{\int_{t_0}^T v^*(t)R(t)v(t)\,dt}{\int_{t_0}^T \{x^*(t)Q(t)x(t) + u^*(t)U(t)u(t)\}\,dt}. \tag{12}$$

Among all $C(t)$ that stabilize (9) and (10), find one that maximizes λ.

2. PROBLEM FORMULATION

The problems considered in Section 1 can be recast in the following form. Consider the time-varying linear system described by

$$\dot{x}_p = F(t)x_p(t) + B_1(t)v(t) + B_2(t)u(t), \quad x_p(t_0) = 0, \tag{13}$$

$$\dot{x}_c = F_c(t)x_c(t) + B_c(t)y(t), \quad x_c(t_0) = 0, \tag{14}$$

$$u = C_c(t)x_c, \tag{15}$$

$$z = C_1(t)x_p(t) + D_1(t)u, \tag{16}$$

$$y = C_2(t)x_p(t) + D_2(t)u, \tag{17}$$

where $x_p(t), u(t), x_c(t), y(t), v(t)$, and $z(t)$ denote the state vector, the control vector, the control state vector, the output vector, the exogenous input vector, and the vector to be controlled respectively. The desired value of $z(t)$ is zero.

There is freedom in the choice of the control parameter matrices $F_c(t), B_c(t)$, and $C_c(t)$. Suppose the quotient

$$\frac{\int_{t_0}^{T} v^*(t)R(t)v(t)\,dt}{\int_{t_0}^{T} z^*(t)Q(t)z(t)\,dt} \tag{18}$$

attains a minimum value λ for some $v_0(t)$ for fixed $F_c(t), B_c(t)$, and $C_c(t)$. Thus $v_0(t)$ denotes the worst normalized disturbance and $1/\lambda$ is a measure of the maximum effect of the disturbance on the system error $z(t)$. In fact it gives the ratio of the weighted error energy to the weighted disturbance energy in the case of the worst possible disturbance for a given set of control parameter matrices.

Now the problem can be stated as follows. Choose one or more of the control parameter matrices $F_c(t), B_c(t)$, and $C_c(t)$ such that

(1) the closed loop system is stable in some sense; and

(2) the value of λ is made as large as possible.

While we establish the existence of λ for given $F_c(t), B_c(t)$, and $C_c(t)$, the question of existence of a maximum value of λ with respect to one or more of the matrices $F_c(t), B_c(t)$, and $C_c(t)$ will not be addressed here. The case in which (15) is of the form $u = C_c(t)x_c + D_c(t)y$ can be handled in an analogous manner.

3. EXISTENCE OF THE WORST EXOGENOUS INPUT

Given $F_c(t), B_c(t)$, and $C_c(t)$, equations (13)-(17) can be rewritten as

$$\begin{pmatrix} \dot{x}_p \\ \dot{x}_c \end{pmatrix} = \begin{pmatrix} F & B_2 C_c \\ B_c C_2 & F_c + B_c D_2 C_c \end{pmatrix} \begin{pmatrix} x_p \\ x_c \end{pmatrix} + \begin{pmatrix} B_1(t) \\ 0 \end{pmatrix} v \tag{19}$$

with

$$z(t) = (C_1(t) \quad D_1(t)C_c(t)) \begin{pmatrix} x_p(t) \\ x_c(t) \end{pmatrix}. \tag{20}$$

Let $x_p(t_0) = 0$ and $x_c(t_0) = 0$. We now consider the conditions under which there exists a worst exogenous input $v_0(t)$ for which

$$\frac{\int_{t_0}^T v^*(t)R(t)v(t)\, dt}{\int_{t_0}^T z^*(t)Q(t)z(t)\, dt} \tag{21}$$

attains a minimum.

In this section $T \le \infty$. We make the following assumptions.

(a) $F(t), B_1(t), B_2(t), C_1(t), C_2(t), D_1(t), D_2(t), F_c(t), B_c(t), C_c(t), R(t)$, and $Q(t)$ are continuous on $[t_0, T]$.

(b) $R(t) > 0$ and $Q(t) \ge 0$.

(c) Admissible exogenous inputs are those for which $\int_{t_0}^T v^*(t)R(t)v(t)\, dt < \infty$. We assume that $v^* R(t)v \ge a\, v^* v$ for some $a > 0$ for all t. Thus each admissible exogenous input belongs to the Hilbert space $L_2(t_0, T)$.

(d) For each $K < \infty$, there is a Lebesgue integrable $g_K(t)$ such that if $\|v\|_2 \leq K$, then $z^*(t)Q(t)z(t) \leq g_K(t)$ almost everywhere on $[t_0, T]$.

(e) There exists an exogenous input for which $0 < \int_{t_0}^T z^*(t)Q(t)z(t)\, dt < \infty$.

Assumption (a) implies that for every fixed $t \in [t_0, T)$, $z(t)$ is a linear continuous vector functional of $v \in L_2(t_0, T)$. Assumptions (a) and (b) imply that for each $T < \infty$, $v^*(t)R(t)v(t) \leq k\, v^*(t)v(t)$ for all $t \in [t_0, T]$ for some $k < \infty$. Thus, if $v \in L_2(t_0, T)$, then it is admissible. Moreover, $v^*R(t)v \geq a\, v^*v$ for some $a > 0$ for all t (assumption (c)) implies that all admissible exogenous inputs belong to $L_2(t_0, T)$. Although the following theorem follows from Theorem 2.1 of Chapter 2, we give a proof to make this chapter as self-contained as possible.

THEOREM 3.1. *Consider the system given by (19)-(21) along with assumptions (a)-(e). Then there exists an admissible exogenous input which minimizes (21).*

Proof. Let λ denote the infimum of (21), which exists by assumption (e). It suffices to consider only those exogenous inputs for which $\int_{t_0}^T z^*(t)Q(t)z(t)\, dt = 1$ since the cost is invariant under exogenous input scaling. Let $\{v_i\}$ be an admissible sequence such that $\lim_{i \to \infty} \int_{t_0}^T v_i^*(t)R(t)v_i(t)\, dt = \lambda$ with $\int_{t_0}^T z_i^*(t)Q(t)z_i(t)\, dt = 1$ for each i.

Since $\{v_i\}$ is bounded in $L_2(t_0, T)$, a subsequence, still denoted by $\{v_i\}$ converges weakly to some $v_0 \in L_2(t_0, T)$. If $T < \infty$, since v^*Rv is convex in v,

$$\int_{t_0}^T \{v_i^*Rv_i - v_0^*Rv_0\}\, dt \geq 2\int_{t_0}^T v_0^*R(v_i - v_0)\, dt.$$

As $i \to \infty$, the right side of the above inequality goes to zero as a consequence of weak convergence. Thus

$$\int_{t_0}^T v_0^*(t)R(t)v_0(t)\, dt \leq \liminf_{i \to \infty} \int_{t_0}^T v_i^*(t)R(t)v_i(t)\, dt = \lambda.$$

If $T = \infty$, let $[t_0, T) = \bigcup_{j=0}^{\infty}[t_0 + j, t_0 + j + 1]$. We have

$$a_j = \int_{t_0+j}^{t_0+j+1} v_0^*(t)R(t)v_0(t)\, dt$$

$$\leq \liminf_{i\to\infty} \int_{t_0+j}^{t_0+j+1} v_i^*(t)R(t)v_i(t)\, dt.$$

So by a discrete version of Fatou's lemma

$$\sum_{j=0}^{\infty} a_j \leq \sum_{j=0}^{\infty} \liminf_{i\to\infty} \int_{t_0+j}^{t_0+j+1} v_i^*(t)R(t)v_i(t)\, dt$$

$$\leq \liminf_{i\to\infty} \sum_{j=0}^{\infty} \int_{t_0+j}^{t_0+j+1} v_i^*(t)R(t)v_i(t)\, dt$$

$$= \lambda.$$

Hence $\int_{t_0}^{T} v_0^*(t)R(t)v_0(t)\, dt = \lambda$ for $T = \infty$. By the weak convergence of $\{v_i\}$, since $z(t)$ is a linear continuous vector functional of $v \in L_2(t_0, T)$ for each fixed $t \in [t_0, T)$, $z_i(t)$ converges pointwise to $z_0(t)$, which is the error response corresponding to $v_0(t)$. Since $\{v_i\}$ is bounded in $L_2(t_0, T)$, by assumption (d) and by the Lebesgue dominated convergence theorem, we conclude that

$$\int_{t_0}^{T} z_0^*(t)Q(t)z_0(t)\, dt = 1.$$

Thus the proof is complete.   □

4. EVALUATION OF λ

In this section, we assume that $T < \infty$. Supposing that $F_c(t), B_c(t)$, and $C_c(t)$ stabilize the system, we give a criterion to evaluate λ in a general case by converting the minimization problem into a boundary value problem. This general case encompasses the class of problems considered in Section 3. It also covers cases not considered in Section 3, such as the inclusion of terms containing $v(t)$ in the denominator of the performance index. If the stability assumption mentioned above is satisfied, then assumption (d) of Section 3 is satisfied. Now, let $x = (x_p^*, x_c^*)^*$ and write (19) and (20) as

$$\dot{x} = A(t, \theta(t))x + B(t, \theta(t))v, \quad x(t_0) = d, \quad x(T) \text{ free,} \tag{22}$$

and the cost functional as

$$\frac{\int_{t_0}^T \frac{1}{2} v^*(t) R_3(t) v(t)\, dt}{\int_{t_0}^T \{\frac{1}{2} x^*(t) W_1(t) x(t) + x^*(t) W_2(t) v(t) + \frac{1}{2} v^*(t) W_3(t) v(t)\}\, dt}, \tag{23}$$

where $\theta(t)$ is a parametrization of the control function. For example, $\theta(t)$ may include the matrix functions $F_c(t), B_c(t), C_c(t)$, and $D_c(t)$ which characterize $u(t)$ in Section 2. To be able to apply Theorem 3.1 of Section 3 to the cost functional (23), we need to set $W_2(t) = W_3(t) = 0$ and $x(t_0) = 0$. Note that $R_3(t) > 0, W_1(t) \geq 0$, and $W_3(t) \geq 0$. Although it is not shown explicitly, $W_1(t), W_2(t)$, and $W_3(t)$ may vary with $\theta(t)$. Assume that $A(t), B(t), R_3(t), W_1(t), W_2(t)$, and $W_3(t)$ are continuous. An exogenous input $v(t)$ is admissible if and only if $v(t) \in L_2(t_0, T)$. In order for the minimization problem to be nontrivial, we assume that there is an admissible $v(t)$ for which the denominator of (23) is positive. We now find conditions that are satisfied by an optimal $v_0(t) \in L_2(t_0, T)$ which minimizes (23) subject to (22).

THEOREM 4.1. *Consider the system given by (22) and (23). If $(x_0(t), v_0(t))$ is optimal, then there exists a nonzero $\psi(t)$ such that*

$$\frac{d\psi}{dt} = -[A + \lambda B(R_3 - \lambda W_3)^{-1} W_2^*]^* \psi$$

$$-[\lambda W_1 + \lambda^2 W_2(R_3 - \lambda W_3)^{-1} W_2^*] x_0, \quad \psi(T) = 0, \tag{24}$$

where

$$\lambda = \inf_v \frac{\int_{t_0}^T \frac{1}{2} v^* R_3 v\, dt}{\int_{t_0}^T \{\frac{1}{2} x^* W_1 x + x^* W_2 v + \frac{1}{2} v^* W_3 v\}\, dt}, \tag{25}$$

and

$$v_0(t) = (R_3 - \lambda W_3)^{-1}\{B^*\psi + \lambda W_2^* x_0\}. \tag{26}$$

Proof. If $v_0(t)$ minimizes (23), then it also minimizes

$$J(v) \triangleq \int_{t_0}^T [\frac{1}{2}v^* R_3 v - \lambda\{\frac{1}{2}x^* W_1 x + x^* W_2 v + \frac{1}{2}v^* W_3 v\}] \, dt. \tag{27}$$

By the maximal principle [9], there exists an adjoint response $\psi(t)$ such that the Hamiltonian

$$H(\psi, x, v) = -\frac{1}{2}v^*(R_3 - \lambda W_3)v + \frac{1}{2}\lambda x^* W_1 x + \lambda x^* W_2 v$$

$$+\psi^*(t)(A(t)x(t) + B(t)v(t)) \tag{28}$$

is maximized almost everywhere on $[t_0, T]$ by $v_0(t)$. Satisfaction of $\partial H/\partial v = 0$ yields

$$v_0(t) = (R_3 - \lambda W_3)^{-1}\{B^*\psi + \lambda W_2^* x_0\}. \tag{29}$$

The adjoint variable $\psi(t)$ satisfies

$$\frac{d\psi}{dt} = -\frac{\partial H}{\partial x}(\psi, x_0, v_0). \tag{30}$$

Thus we have

$$\frac{d\psi}{dt} = -[A + \lambda B(R_3 - \lambda W_3)^{-1}W_2^*]^*\psi - [\lambda W_1 + \lambda^2 W_2(R_3 - \lambda W_3)^{-1}W_2^*]x_0. \tag{31}$$

The transversality condition yields $\psi(T) = 0$.   □

Let

$$\hat{A} = A + \lambda B(R_3 - \lambda W_3)^{-1}W_2^*,$$

$$\hat{B} = B(R_3 - \lambda W_3)^{-1}B^*, \tag{32}$$

$$\hat{C} = -\lambda W_1 - \lambda^2 W_2(R_3 - \lambda W_3)^{-1}W_2^*.$$

Thus we have a two-point boundary value problem given by

$$\begin{pmatrix} \dot{x}_0 \\ \dot{\psi} \end{pmatrix} = \begin{pmatrix} \hat{A} & \hat{B} \\ \hat{C} & -\hat{A}^* \end{pmatrix} \begin{pmatrix} x_0 \\ \psi \end{pmatrix} \tag{33}$$

with

$$x_0(t_0) = d, \quad \psi(T) = 0. \tag{34}$$

We now give a criterion for the estimation of λ.

THEOREM 4.2. *Consider the system*

$$\dot{x} = A(t, \theta(t))x + B(t, \theta(t))v, \quad x(t_0) = 0, \quad x(T) \text{ is free}, \tag{35}$$

and assume that there exists an exogenous input which minimizes (23). Now consider the boundary value problem given by

$$\begin{pmatrix} \dot{x} \\ \dot{\psi} \end{pmatrix} = \begin{pmatrix} \hat{A} & \hat{B} \\ \hat{C} & -\hat{A}^* \end{pmatrix} \begin{pmatrix} x \\ \psi \end{pmatrix}, \tag{36}$$

$$\begin{pmatrix} x(t_0) \\ \psi(T) \end{pmatrix} = \begin{pmatrix} 0 \\ 0 \end{pmatrix}, \tag{37}$$

where $\hat{A}, \hat{B},$ and $\hat{C}$ are as defined by (32). Note that λ is a parameter in $\hat{A}, \hat{B},$ and $\hat{C}$. Let $v \triangleq (R_3 - \lambda W_3)^{-1}\{B^*\psi + \lambda W_2^* x\}$. If λ is the smallest positive number such that (36)-(37) has a solution $(x_0(t), \psi(t))$ with $\int_{t_0}^{T}\{\frac{1}{2}x_0^* W_1 x_0 + x_0^* W_2 v_0 + \frac{1}{2}v_0^* W_3 v_0\}\, dt > 0$, then λ is the optimal value. Moreover, x_0 is an optimal trajectory and $v_0 = (R_3 - \lambda W_3)^{-1}\{B^*\psi + \lambda W_2^* x_0\}$ is an optimal exogenous input.

Proof. It is clear from Theorem 4.1 that if $x_0(t)$ minimizes (23), it satisfies (36) and (37), with λ being the minimum value of (23). Now suppose (x, ψ) is a solution of (36) and (37) for some λ.

Let $\Lambda = (R_3 - \lambda W_3)^{-1}$ and $v = \Lambda\{B^*\psi + \lambda W_2^* x\}$. We have

$$\int_{t_0}^{T} v^*(R_3 - \lambda W_3)v\, dt = \int_{t_0}^{T} (\Lambda\{B^*\psi + \lambda W_2^* x\}, B^*\psi + \lambda W_2^* x)\, dt$$

$$= \int_{t_0}^{T} (B\Lambda\{B^*\psi + \lambda W_2^* x\}, \psi)\, dt + \lambda \int_{t_0}^{T} x^* W_2 v\, dt$$

$$= \int_{t_0}^{T} (\dot{x}, \psi)\, dt - \int_{t_0}^{T} (Ax, \psi)\, dt + \lambda \int_{t_0}^{T} x^* W_2 v\, dt. \qquad (38)$$

Integrating the first integral in (38) by parts, making use of $x(t_0) = \psi(T) = 0$, and rearranging, we get

$$\int_{t_0}^{T} v^* R_3 v\, dt = \lambda \int_{t_0}^{T} \{(x, W_1 x) + 2(x, W_2 v) + (v, W_3 v)\}\, dt. \qquad (39)$$

Thus, if $(x_0(t), \psi(t))$ is a solution of the boundary value problem given by (36) and (37) for the smallest parameter $\lambda > 0$ with $\int_{t_0}^{T}\{x_0^* W_1 x_0 + 2x_0^* W_2 v_0 + v_0^* W_3 v_0\}\, dt > 0$, then $x_0(t)$ is optimal. $\square$

Note that the boundary value problem (36)-(37) has a solution with a nonvanishing denominator for (23) for at most a countably infinite values of λ. Theorem 4.2 gives a sufficient condition for an exogenous input to be optimal. Thus Theorem 4.1 (with $x(t_0) = 0$) and Theorem 4.2 give a complete characterization of an optimal exogenous input.

Making use of the transition matrix, the solution of (36) may be expressed as

$$\begin{pmatrix} x(t) \\ \psi(t) \end{pmatrix} = \begin{pmatrix} \Phi_{11}(t, t_0) & \Phi_{12}(t, t_0) \\ \Phi_{21}(t, t_0) & \Phi_{22}(t, t_0) \end{pmatrix} \begin{pmatrix} x(t_0) \\ \psi(t_0) \end{pmatrix}. \qquad (40)$$

Equation (37) yields

$$\Phi_{12}(T, t_0)\psi(t_0) = x(T), \qquad (41)$$

$$\Phi_{22}(T, t_0)\psi(t_0) = 0. \qquad (42)$$

In view of (42) and (36)-(37), we have $\det(\Phi_{22}(T, t_0)) = 0$ if and only if the solution (x_0, ψ) of (36)-(37) is not identically zero. Thus, we need the least positive λ which makes $\det(\Phi_{22}(T, t_0)) = 0$ and the denominator of (23) positive. This can be usually obtained by doing a search with λ over an interval on which there is a change in the sign of the determinant.

We found the following algorithm to be numerically more stable since numbers of lesser magnitude are involved in the computation of the transition matrices in (43). We have

$$\begin{pmatrix} x(T) \\ \psi(T) \end{pmatrix} = \Phi(T, \frac{T + t_0}{2})\Phi(\frac{T + t_0}{2}, t_0) \begin{pmatrix} x(t_0) \\ \psi(t_0) \end{pmatrix}. \tag{43}$$

Let

$$\Phi^{-1}(T, \frac{T + t_0}{2}) = \begin{pmatrix} \zeta_{11} & \zeta_{12} \\ \zeta_{21} & \zeta_{22} \end{pmatrix}$$

and

$$\Phi(\frac{T + t_0}{2}, t_0) = \begin{pmatrix} \nu_{11} & \nu_{12} \\ \nu_{21} & \nu_{22} \end{pmatrix}.$$

Making use of $x(t_0) = \psi(T) = 0$, we have

$$\nu_{12}\psi(t_0) = \zeta_{11}x(T), \tag{44}$$

$$\nu_{22}\psi(t_0) = \zeta_{21}x(T). \tag{45}$$

Thus

$$\det \begin{pmatrix} \nu_{12} & \zeta_{11} \\ \nu_{22} & \zeta_{21} \end{pmatrix} = 0. \tag{46}$$

Thus we need the least positive λ which makes the above determinant zero.

5. SOME APPLICATIONS

In the case of a full-order observer, the equations are given by

$$\dot{x}_p = F(t)x_p + B_1(t)v + B_2(t)u, \tag{47}$$

$$\dot{x}_c = (F(t) - L(t)C_2(t))x_c + B_2(t)u + L(t)C_2(t)x_p, \tag{48}$$

where x_p and x_c have the same dimension. Assume that $L(t)$ is known. Letting $u(t) = C_c(t)x_c(t)$ and $z(t) = (x_p^*(t), u^*(t), v^*(t))^*$, we get

$$\begin{pmatrix} \dot{x}_p \\ \dot{x}_c \end{pmatrix} = \begin{pmatrix} F & B_2C_c \\ LC_2 & F - LC_2 + B_2C_c \end{pmatrix} \begin{pmatrix} x_p \\ x_c \end{pmatrix} + \begin{pmatrix} B_1(t) \\ 0 \end{pmatrix} v. \tag{49}$$

Assuming the initial conditions to be zero, the problem is to choose $C_c(t)$ such that (49) is stable and the minimum value of

$$\frac{\int_{t_0}^{T} v^*(t)R(t)v(t)\, dt}{\int_{t_0}^{T} z^*(t)Q(t)z(t)\, dt} \tag{50}$$

is maximized.

For time-invariant systems with $t_0 = 0$ and $T = \infty$, (13)-(17) may be expressed in terms of Laplace transforms as

$$Z(s) = G(s)V(s). \tag{51}$$

Using Plancherel's theorem and assuming $R(t) = Q(t) = I$, (18) may be written as

$$\frac{\int_{-\infty}^{\infty} V^*(j\omega)V(j\omega)\, d\omega}{\int_{-\infty}^{\infty} Z^*(j\omega)Z(j\omega)\, d\omega}, \tag{52}$$

where the superscript $*$ here denotes complex-conjugate transpose. Since the infimum of (52) is the reciprocal of the square of the H_∞-norm of $G(s)$, the problem in Section 2 is reduced to choosing the matrices F_c, B_c, and C_c such that the system is stable and the H_∞-norm of $G(s)$ is minimized.

The design procedure given in this chapter may be summarized for time-invariant systems as follows. Consider (13)-(18). Assuming values to the matrices F_c, B_c, and C_c to make the closed loop system stable, find λ. Iterate on the elements of one or more matrices using an optimization routine to maximize λ while maintaining the stability of the system.

For time-varying systems, a similar procedure can be employed by expanding the elements of the matrices in terms of basis functions and maximizing λ with respect to the coefficients of these basis functions. The existence of a maximum λ with respect to these coefficients is not considered here.

As an illustration of the concepts involved, we now present a simple scalar example. A multivariable application will be presented in Section 7. The system is described by the equation

$$\dot{x} = -x + u + v, \quad x(0) = 0, \quad u = cx, \tag{53}$$

and the objective is to choose c which stabilizes the system and maximizes the minimum of

$$\frac{\int_0^1 v^2(t)\,dt}{\int_0^1 (x^2 + u^2)\,dt} = \frac{\int_0^1 v^2(t)\,dt}{(1 + c^2)\int_0^1 x^2\,dt}. \tag{54}$$

From equations (33) and (34), we get

$$\begin{pmatrix} \dot{x} \\ \dot{\psi} \end{pmatrix} = \begin{pmatrix} c - 1 & 1 \\ -\lambda(1 + c^2) & -(c - 1) \end{pmatrix} \begin{pmatrix} x \\ \psi \end{pmatrix} \tag{55}$$

with

$$x(0) = 0, \quad \psi(1) = 0. \tag{56}$$

Let $a = 1 - c$ be fixed, which needs to be positive for stability. From the material of Section 3, there exists a $v_0(t)$ which minimizes (54). According to the theory of Section 4, we need to find the least positive λ which makes the element $\Phi_{22}(t)$ of the transition matrix of (55) vanish at $t = 1$.

In terms of Laplace transforms, we have

$$\Phi_{22}(s) = \frac{s + a}{s^2 + \lambda(1 + c^2) - a^2}. \tag{57}$$

Case 1. $\lambda < \dfrac{a^2}{1 + c^2}$

Let $\lambda(1 + c^2) - a^2 = -d^2, d > 0$. In this case the requirement that $\Phi_{22}(1) = 0$ leads to the equation

$$\frac{a - d}{a + d} = \exp(2d), \tag{58}$$

which does not have a solution in $d \in (0, \infty)$.

Case 2. $\lambda = \dfrac{a^2}{1 + c^2}$

In this case $\Phi_{22}(1) = 0$ implies that $a = -1$, which is not allowed.

Case 3. $\lambda > \dfrac{a^2}{1 + c^2}$

Let $\lambda(1 + c^2) - a^2 = d^2, d > 0$. The condition $\Phi_{22}(1) = 0$ implies that

$$\tan d = -\frac{d}{a} = \frac{d}{1 - c}. \tag{59}$$

The above equation has countably infinite solutions in d and we find the least positive solution. Then

$$\lambda = \frac{a^2 + d^2}{1 + c^2} = \frac{(1 - c)^2 + d^2}{1 + c^2}. \tag{60}$$

Since d which is a function of c is bounded above by π, due to (59) $\lambda \to 1$ as $|c| \to \infty$. The value of $c < 1$ which maximizes λ can be found using an optimization routine. The optimal value is $c = -0.340$ with $\lambda_{max} = 5.68371$.

For higher dimensional problems, the design procedure requires the use of a digital computer. We give a higher dimensional example in Section 7.

6. PERFORMANCE ROBUSTNESS

In this section we develop a formula for the variation of λ when there are parameter variations in the system matrices. Our state space formulation is convenient to handle parameter uncertainties. For this consider

$$\dot{x} = A(t, \theta(t))x + B(t, \theta(t))v, \quad x(t_0) = 0, \quad x(T) \text{ free}, \tag{61}$$

with the performance index

$$J(v) = \frac{\int_{t_0}^{T} \frac{1}{2} v^* R_3 v \, dt}{\int_{t_0}^{T} \{\frac{1}{2} x^* W_1 x + x^* W_2 v + \frac{1}{2} v^* W_3 v\} \, dt}. \tag{62}$$

In the following analysis, we only consider variations in the matrices A and B. If there is an output equation which is subject to variations also, the analysis can be readily extended, taking into account the fact that there may be corresponding variations in the weighting matrices W_1, W_2, and W_3.

Let a controller be characterized by $\theta(t)$ and λ be its performance measure, which is the infimum of (62) over $v(t)$. Let μ denote the variation in λ for elemental variations δA and δB in A and B. For performance robustness, we require that

$$|\mu/\lambda| \le \mu_0 \text{ for all } \|\delta A(t)\| \le a(t) \text{ and } \|\delta B(t)\| \le b(t). \tag{63}$$

We now state the performance robustness problem.

Performance Robustness Problem. Select a controller characterization $\theta(t)$ such that

$$\inf_v \frac{\int_{t_0}^{T} \frac{1}{2} v^* R_3 v \, dt}{\int_{t_0}^{T} \{\frac{1}{2} x^* W_1 x + x^* W_2 v + \frac{1}{2} v^* W_3 v\} \, dt} \tag{64}$$

is maximized with the side constraint

$$|\mu/\lambda| \le \mu_0 \text{ for all } \|\delta A(t)\| \le a(t) \text{ and } \|\delta B(t)\| \le b(t).$$

We now derive an expression for μ in terms of variations in the system matrices. For a given characterization $\theta(t)$ of the control, let $v(t)$ minimize (62). Let

$$\Lambda = (R_3 - \lambda W_3)^{-1},$$

$$\hat{A} = A + \lambda B\Lambda W_2^*,$$

$$\hat{B} = B\Lambda B^*, \tag{65}$$

$$\hat{C} = -\lambda W_1 - \lambda^2 W_2 \Lambda W_2^*.$$

From the theory of Section 4, we have a boundary value problem given by

$$\dot{x} = \hat{A}x + \hat{B}\psi, \tag{66}$$

$$\dot{\psi} = \hat{C}x - \hat{A}^*\psi, \tag{67}$$

with

$$x(t_0) = 0, \quad \psi(T) = 0. \tag{68}$$

Let x_1 and ψ_1 represent variations in x and ψ owing to elemental variations δA and δB. Let the corresponding variation in λ be denoted by μ. We have the following set of equations that are satisfied by x_1 and ψ_1:

$$\dot{x}_1 = \hat{A}x_1 + \hat{B}\psi_1 + (P_1 + \mu Q_1)x + (P_2 + \mu Q_2)\psi, \tag{69}$$

$$\dot{\psi}_1 = \hat{C}x_1 - \hat{A}^*\psi_1 + \mu Q_3 x - (P_1 + \mu Q_1)^*\psi, \tag{70}$$

$$x_1(t_0) = 0, \quad \psi_1(T) = 0, \tag{71}$$

where

$$P_1 = \delta A + \lambda\, \delta B\, \Lambda W_2^*,$$

$$P_2 = \delta B\, \Lambda B^* + B\Lambda\, \delta B^*,$$

$$Q_1 = \lambda B\Lambda W_3\Lambda W_2^* + B\Lambda W_2^*, \tag{72}$$

$$Q_2 = B\Lambda W_3\Lambda B^*,$$

$$Q_3 = -W_1 - \lambda^2 W_2\Lambda W_3\Lambda W_2^* - 2\lambda W_2\Lambda W_2^*.$$

THEOREM 6.1. *The variation μ in performance is given by*

$$\mu = \frac{- \int_{t_0}^{T} \{x^* P_1^* \psi + \frac{1}{2} \psi^* P_2 \psi\} \, dt}{\int_{t_0}^{T} \{\frac{1}{2} x^* W_1 x + x^* W_2 v + \frac{1}{2} v^* W_3 v\} \, dt}. \tag{73}$$

In order to prove the above theorem, we need a preliminary lemma.

LEMMA 6.1. *Let $v = \Lambda\{B^* \psi + \lambda W_2^* x\}$. Then*

$$\int_{t_0}^{T} \{x^* Q_3 x - 2 x^* Q_1^* \psi - \psi^* Q_2 \psi\} \, dt = - \int_{t_0}^{T} \{x^* W_1 x + 2 x^* W_2 v + v^* W_3 v\} \, dt. \tag{74}$$

Proof. By substituting the expressions for Q_1, Q_2, and Q_3 from equation (72) and grouping terms, we can show that the left side of (74) equals

$$\int_{t_0}^{T} \{-(\psi^* B + \lambda x^* W_2)\Lambda W_3 \Lambda B^* \psi - \lambda x^* W_2 \Lambda W_3 \Lambda(B^* \psi + \lambda W_2^* x)$$
$$-2 x^* W_2 \Lambda(B^* \psi + \lambda W_2^* x)\} \, dt.$$

The above expression can be easily shown to be equal to the right side of (74).   □

We now prove Theorem 6.1.

Proof of Theorem 6.1. From (70) we get

$$\int_{t_0}^{T} x^* \dot{\psi}_1 \, dt = \int_{t_0}^{T} x^* \hat{C} x_1 \, dt - \int_{t_0}^{T} x^* \hat{A}^* \psi_1 \, dt$$
$$+ \mu \int_{t_0}^{T} x^* Q_3 x \, dt - \int_{t_0}^{T} x^* (P_1 + \mu Q_1)^* \psi \, dt. \tag{75}$$

Integrating the left side of (75) by parts and making use of (66), we get

$$- \int_{t_0}^{T} \psi_1^* \hat{B} \psi \, dt = \int_{t_0}^{T} x^* \hat{C} x_1 \, dt + \mu \int_{t_0}^{T} x^* Q_3 x \, dt$$
$$- \int_{t_0}^{T} x^* (P_1 + \mu Q_1)^* \psi \, dt. \tag{76}$$

By equation (67) the first integral on the right side of (76) is written as

$$\int_{t_0}^{T} x^* \hat{C} x_1 \, dt = \int_{t_0}^{T} (\dot{\psi} + \hat{A}^* \psi)^* x_1 \, dt. \qquad (77)$$

An integration by parts and equation (69) yield

$$\int_{t_0}^{T} x^* \hat{C} x_1 \, dt = - \int_{t_0}^{T} \psi^* \hat{B} \psi_1 \, dt - \int_{t_0}^{T} \psi^* (P_1 + \mu Q_1) x \, dt$$

$$- \int_{t_0}^{T} \psi^* (P_2 + \mu Q_2) x \, dt. \qquad (78)$$

Substituting (78) in (76) and simplifying, we get

$$\mu = \frac{2 \int_{t_0}^{T} x^* P_1^* \psi \, dt + \int_{t_0}^{T} \psi^* P_2 \psi \, dt}{\int_{t_0}^{T} x^* Q_3 x \, dt - 2 \int_{t_0}^{T} x^* Q_1^* \psi \, dt - \int_{t_0}^{T} \psi^* Q_2 \psi \, dt}. \qquad (79)$$

From Lemma 6.1, the conclusion of Theorem 6.1 follows.   □

Using (73), the variation of performance owing to parameter variations can be computed.

7. NAVION WING LEVELER SYSTEM

We now give a multivariable example. Although it is a time-invariant example, it illustrates the basic methodology.

The lateral dynamics of Navion, a single-engine, four-passenger general aviation aircraft at sea level and at an air speed of 144 ft/sec is given by [10]

$$\dot{x} = Ax + Bu, \qquad (80)$$

where

$$A = \begin{pmatrix} -7.241 & 0 & -6.247 & 1.556 \\ 1 & 0 & 0 & .105 \\ .1056 & .222 & -.221 & -1 \\ -.797 & 0 & 3.743 & -.651 \end{pmatrix}, \qquad (81)$$

$$B = \begin{pmatrix} 18.86 & -2.71 \\ 0 & 0 \\ 0 & -.043 \\ -.07 & 3.51 \end{pmatrix}, \tag{82}$$

$$x = (p \quad \phi \quad \beta \quad r)^*, \tag{83}$$

and

$$u = (\delta_A \quad \delta_R)^*. \tag{84}$$

The states in (83) are roll rate, roll angle, sideslip angle, and yaw rate respectively. In (84) δ_A represents the aileron deflection and δ_R, the rudder deflection. The units for the angles are radians.

We adjoin (80) with a disturbance term and write it as

$$\dot{x} = Ax + Bu + Bv, \qquad x(0) = 0, \tag{85}$$

with

$$u = Cx. \tag{86}$$

The objective is to determine the 2×4 matrix C such that $A + BC$ is stable and the minimum of

$$\frac{\int_0^5 v^* R v \, dt}{\int_0^5 \{x^* Q x + u^* U u\} \, dt} \tag{87}$$

is maximized. In (87) the weighting matrices are taken to be

$$R = \begin{pmatrix} 100 & 0 \\ 0 & 100 \end{pmatrix}, \quad Q = \begin{pmatrix} 1 & 0 & 0 & 0 \\ 0 & 1 & 0 & 0 \\ 0 & 0 & 10 & 0 \\ 0 & 0 & 0 & 5 \end{pmatrix}, \quad U = \begin{pmatrix} 4 & 0 \\ 0 & 4 \end{pmatrix}. \tag{88}$$

Let $W = Q + C^* U C$. Assuming a fixed value for C, according to Section 4, we need to determine the least positive λ for which the boundary value problem given by

$$\begin{pmatrix} \dot{x} \\ \dot{\psi} \end{pmatrix} = \begin{pmatrix} A + BC & BR^{-1}B^* \\ -\lambda W & -(A + BC)^* \end{pmatrix} \begin{pmatrix} x \\ \psi \end{pmatrix}, \tag{89}$$

$$x(0) = 0, \qquad \psi(5) = 0, \tag{90}$$

has a nontrivial solution.

Let $\Phi(t) = \begin{pmatrix} \Phi_{11} & \Phi_{12} \\ \Phi_{21} & \Phi_{22} \end{pmatrix}$ be the 8×8 transition matrix corresponding to (89). Satis-

faction of equation (37) gives rise to the condition that $\det(\Phi_{22}(5)) = 0$. Thus λ is found

by making use of a sign change of $\det(\Phi_{22}(5))$ over a range of values of λ.

The transition matrix $\Phi(5)$ was found by using a Fortran version of the PC-MATLAB

matrix exponential routine. The algorithm consists of scaling the argument matrix by a

power of 2 until its norm is less than $\frac{1}{2}$, evaluating the exponential of the scaled matrix very

accurately by a Padé approximation, and then undoing the scaling by repeated squaring.

The algorithm is essentially Algorithm 11.3-1 of [11].

Initially C was chosen to be

$$\begin{pmatrix} -.3 & -.7 & -.8 & .6 \\ 0 & 0 & 0 & 0 \end{pmatrix}.$$

Using the Rosenbrock hill-climbing algorithm [12], the elements of C were varied to maxi-

mize λ. The constraint that $A + BC$ be stable was not introduced since the unconstrained

run resulted in a stable closed loop system. An initial value for λ of 0.1 was chosen and λ

was incremented in steps of 0.2 until $\det(\Phi(5))$ changed sign. For the next iteration, the

initial value of λ was again taken to be 0.1 and λ was incremented until sign change was

observed in the determinant. This process was repeated until convergence was obtained.

The computations were performed on a Cyber 170-875 computer in Fortran Version

5 with double precision. A local maximum of 22.1 was obtained for

$$C = \begin{pmatrix} -.817 & -1.250 & -1.303 & .733 \\ -.426 & .112 & -.205 & -2.426 \end{pmatrix}. \tag{91}$$

The corresponding eigenvalues of $A + BC$ are $-.5035, -1.1887$, and $-14.6163 \pm i4.5148$.

It was observed that the effect of round-off errors can be significant in the computation of λ for a fixed C. Thus, on a computer with less significant bits, the final time needs to be reduced to obtain a meaningful solution. This problem was not encountered in the method given by (43)-(46). We now give the implementation of this method.

Let $\Phi(t)$ be the transition matrix corresponding to (89) with $\Phi(0) = I$. Let

$$\Phi(2.5) = \begin{pmatrix} \nu_{11} & \nu_{12} \\ \nu_{21} & \nu_{22} \end{pmatrix}$$

and

$$\Phi(-2.5) = \begin{pmatrix} \zeta_{11} & \zeta_{12} \\ \zeta_{21} & \zeta_{22} \end{pmatrix}.$$

From equation (43), we have

$$\Phi(-2.5) \begin{pmatrix} x(5) \\ \psi(5) \end{pmatrix} = \Phi(2.5) \begin{pmatrix} x(0) \\ \psi(0) \end{pmatrix}.$$

Making use of $\psi(5) = x(0) = 0$, we need to find the least positive λ which makes

$$\det \begin{pmatrix} \nu_{12} & \zeta_{11} \\ \nu_{22} & \zeta_{21} \end{pmatrix} = 0.$$

The initial values of C and λ, and the increments for λ from the previous run on the Cyber were retained in this case. The same procedure as in the previous case was followed.

We were able to run the program in Fortran on a Zenith Z-248 personal computer in double precision using the Microsoft Optimizing Compiler Version 4.01. A local maximum of $\lambda = 20.3$ was obtained for

$$C = \begin{pmatrix} -.612 & -.871 & -.965 & .648 \\ -.06 & .049 & -.099 & -1.305 \end{pmatrix}. \tag{92}$$

The corresponding eigenvalues of $A + BC$ are $-.6559, -1.2158, -6.2685$, and -15.9742.

The difference in the values of C for the two runs might be due to the existence of several local maxima in the vicinity of the initial point, the difference in precision of the two computers and the differences in the compilers. The convergence proceeded along different paths on the computers. When C given by (91) was inputted to the Z-248 personal computer, the value of λ was observed to be 22.1, confirming the accuracy of (91). Additional comments on the numerical method can be found in the preface of this monograph.

To get variations in λ owing to variations in the system matrices A and B, equation (73) can be used.

8. Conclusions

In this chapter we presented a method for selecting control parameters such that the effect of disturbance inputs on the variables to be controlled is minimized. For a given set of control parameters, we give a criterion for the evaluation of the least positive value of a parameter occurring in a boundary value problem, whose reciprocal gives the effect of the worst disturbance on the variables to be controlled. The least positive value of the parameter also gives a measure of the performance of the given controller. Further research needs to be done to devise a suitable and efficient method for the evaluation of this value. The least value needs to be maximized with respect to the control parameters in order to maximize the disturbance rejection capacity of the system. Also an expression for the variation of the performance of the controller owing to variations in the system matrices is derived. Although the results are presented from the disturbance rejection point of view, the theory can be applied to command following as well. Using the theory, a controller is designed for the wing leveler system of a small aircraft.

REFERENCES

[1] M. B. SUBRAHMANYAM, On applications of control theory to integral inequalities, *SIAM J. Contr. Optimiz.* **19**, 1981, pp. 479-489.

[2] _____ , On integral inequalities associated with a linear operator equation, *Proc. Amer. Math. Soc.* **92**, 1984, pp. 342-346.

[3] _____ , Necessary conditions for minimum in problems with nonstandard cost functionals, *J. Math. Anal. Appl.* **60**, 1977, pp. 601-616.

[4] _____ , On applications of control theory to integral inequalities, *J. Math. Anal. Appl.* **77**, 1980, pp. 47-59.

[5] _____ , A control problem with application to integral inequalities, *J. Math. Anal. Appl.* **81**, 1981, pp. 346-355.

[6] _____ , An extremal problem for convolution inequalities, *J. Math. Anal. Appl.* **87**, 1982, pp. 509-516.

[7] B. A. FRANCIS AND J. C. DOYLE, Linear control theory with an H_∞ optimality criterion, *SIAM J. Contr. Optimiz.* **25**, 1987, pp. 815-844.

[8] B. A. FRANCIS, "A Course in H_∞ Optimal Control Theory," Lecture Notes in Control and Information Sciences, Vol. 88, Berlin, Springer-Verlag, New York, 1987.

[9] E. B. LEE AND L. MARKUS, "Foundations of Optimal Control Theory," Wiley, New York, 1967.

[10] D. R. DOWNING AND J. R. BROUSSARD, "Digital Flight Control System Analysis and Design," Short Course Notes, University of Kansas, Lawrence, KS.

[11] G. H. GOLUB AND C. F. VAN LOAN, "Matrix Computations," Johns Hopkins, Baltimore, MD, 1983, p. 384.

[12] J. L. KUESTER AND J. H. MIZE, "Optimization Techniques with Fortran," McGraw-Hill, New York, 1973.

CHAPTER 4

Necessary Conditions for Optimal Disturbance Rejection

in Linear Systems

ABSTRACT

Disturbance rejection is an important factor in the synthesis of controllers for several practical systems. In this chapter we derive necessary conditions which need to be satisfied by the controller which yields maximum disturbance rejection. Necessary conditions for maximum disturbance rejection are also derived in the case of an observer-based controller. These conditions are useful in the synthesis of a controller which maximizes the disturbance rejection capacity of the system. The problem considered has connections to the H_∞ control theory. An example is given.

1. INTRODUCTION

Disturbance rejection is an important factor in the design of flight control systems of airplanes for gust load alleviation, especially when the elasticity of the wing needs to be taken into account. Also the effect of disturbances on the response of the airplane needs to be small in several situations, such as an automatic landing and low altitude high speed terrain following. In this chapter we present necessary conditions to be satisfied by a controller which yields optimal disturbance rejection. We derive conditions for the case of a regulator in the first three sections. Necessary conditions are also derived in the case of an observer-based controller.

The problem considered in the first three sections is stated below. The system is described by

$$\dot{x} = A(t)x + B(t)u + G(t)v, \quad x(t_0) = 0, \tag{1}$$

where $x(t), u(t), v(t)$ represent the system state, control, and disturbance input respectively. Let the feed back controller be given by

$$u(t) = C(t)x(t). \tag{2}$$

For a given $C(t)$ in (2), let $v(t)$ be chosen such that the performance index

$$J(v) = \frac{\int_{t_0}^T v^*(t)R(t)v(t)\,dt}{\int_{t_0}^T \{x^*(t)Q_1(t)x(t) + u^*(t)Q_2(t)u(t)\}\,dt} \tag{3}$$

is minimized. Note that the superscript $*$ denotes transpose of a matrix or a vector. In (3), $R(t) > 0, Q_1(t) \geq 0$, and $Q_2(t) \geq 0$. Let $\lambda = \inf_v J(v)$. Thus $v(t)$ which minimizes (3) represents the worst disturbance and λ gives the disturbance rejection capacity of the controller given by (2). The problem is to choose $C(t)$ which stabilizes (1) and minimizes λ.

It can be observed that the proposed problem has connections to the H_∞ optimal control theory. We consider the problem from a different point of view and this will be explained in the next few sections. Also, the results are applicable to time-varying systems over a finite horizon.

We now review the current state of the art in H_∞ control theory and outline the contributions made by this chapter. Most of the problems treated in H_∞ control over the past few years involve time-invariant systems. Ref. 1 treats a specialized problem and gives a parametrization of all stabilizing controllers that achieve a specified H_∞-norm bound. The computation of the controller involves the solution of two Riccati equations. This result has been generalized recently [2]. Ref. 3 makes use of a generalized algebraic operation called conjugation to solve the H_∞ problem and once again the computation of the controller involves the solution of two Riccati equations. In Ref. 4, a certain LQG problem with a side constraint on the H_∞-norm of the closed loop transfer function is solved. In this approach it is necessary to solve three Riccati equations. In special cases, these three equations can be reduced to two Riccati equations. Also, [5] and [6] show that when the measured outputs are the states of the plant, one can choose a constant gain suboptimal controller. A formula for the state-feedback gain matrix is given in terms of an algebraic Riccati equation.

The state space approach taken in this chapter results in a two-point boundary value problem in which λ defined above is a parameter. For the general time-varying system with a given controller, the parameter λ gives a measure of the performance of the controller. The parameter λ is related to γ^{-2} found in Section II of [1]. However, the problem considered in [1] is to find an admissible suboptimal controller with H_∞-norm less than a prescribed value. Tadmor [7] extends these results to the finite-horizon time-varying case by considering a series of LQ-optimization problems. He characterizes suboptimal values and gives a parametrization of all suboptimal controllers in the case of finite-horizon time-varying problems. An expression for a suboptimal controller which satisfies a prescribed norm bound is derived via two dynamic matrix Riccati equations.

Our approach is to evaluate λ for a given controller and iterate on the controller using a nonlinear programming algorithm to maximize λ. Although it is relatively simple to derive conditions which characterize the worst disturbance for a given controller, the characterization of λ as the minimum value of a parameter occurring in a boundary value problem as in Theorem 2.2 is new. Also new are the necessary conditions to be satisfied by a controller which maximizes the performance measure λ. These necessary conditions are given in Theorems 3.1, 6.1, and 7.1. These results have not previously been derived even for the time-invariant case. These conditions are useful to verify that the controller obtained from a nonlinear programming algorithm is indeed optimal. Also, novel techniques associated with boundary value problems are used to derive them. Several nonlinear pro-gramming algorithms make use of gradients with respect to the independent variables. Since λ is to be maximized, new expressions for the variation of λ in terms of variations in the controller have been derived in the static as well as the dynamic controller case. These expressions are especially useful for the time-invariant case. For the time-varying case, the controller needs to be expanded in terms of basis functions and λ needs to be maximized with respect to the coefficients of these basis functions.

Also our current research indicates that parameter variations can be conveniently han-dled in our state space formulation. The important robust performance problem, namely, how to maximize the controller performance with a side constraint on the performance variation owing to parameter variations, is addressed in Chapters 3 and 6. It is possible to extend the results of this chapter to the case in which the integrands of the numerator and denominator of (3) are functions obeying certain homogeneity properties outlined in [8-9]. This extension is the subject matter of Chapter 6.

2. NECESSARY CONDITIONS FOR A FIXED $C(t)$

The equations (1)-(3) become

$$\dot{x} = (A(t) + B(t)C(t))x(t) + G(t)v(t), \quad x(t_0) = 0, \tag{4}$$

and

$$J(v) = \frac{\int_{t_0}^{T} v^*(t)R(t)v(t)\,dt}{\int_{t_0}^{T} x^*(t)\{Q_1(t) + C^*(t)Q_2(t)C(t)\}x(t)\,dt}. \tag{5}$$

The problem in this section is to choose $v(t)$ such that (5) is minimized.

Optimal control problems in which the performance index is of the form of a quotient of definite integrals have been considered in [8-10]. We now state the conditions that are satisfied by $v(t)$ which minimizes (5).

THEOREM 2.1. *Consider the system given by (4) and (5). If $(x_0(t), v_0(t))$ minimizes (5), then there exists a nonzero $\psi_0(t)$ such that*

$$\frac{d\psi_0}{dt} = -(A(t) + B(t)C(t))^*\psi_0(t) - \lambda W(t)x_0(t), \quad \psi_0(T) = 0, \tag{6}$$

where

$$W(t) = Q_1(t) + C^*(t)Q_2(t)C(t), \tag{7}$$

$$\lambda = \inf_v \frac{\int_{t_0}^T v^*(t)R(t)v(t)\, dt}{\int_{t_0}^T x^*(t)W(t)x(t)\, dt}, \tag{8}$$

and

$$v_0(t) = R^{-1}(t)G^*(t)\psi_0(t). \tag{9}$$

Proof. If $v_0(t)$ minimizes (5), then it also minimizes

$$J_1(v) \triangleq \frac{1}{2} \int_{t_0}^T \{v^*(t)R(t)v(t) - \lambda x^*(t)W(t)x(t)\}\, dt. \tag{10}$$

The theorem follows by applying the maximal principle [11] to the performance index given by (10). □

Thus, we have a two-point boundary value problem given by

$$\begin{pmatrix} \dot{x}_0 \\ \dot{\psi}_0 \end{pmatrix} = \begin{pmatrix} A + BC & GR^{-1}G^* \\ -\lambda W(t) & -(A + BC)^* \end{pmatrix} \begin{pmatrix} x_0 \\ \psi_0 \end{pmatrix}, \tag{11}$$

with

$$x_0(t_0) = 0, \quad \psi_0(T) = 0. \tag{12}$$

It is shown below that the minimum value of (5) is given by the smallest positive value of λ for which the boundary value problem given by (11) and (12) has a solution with $\int_{t_0}^T x^*(t)W(t)x(t)\, dt > 0$.

THEOREM 2.2. *Let (x, ψ) satisfy the bounadry value problem*

$$\begin{pmatrix} \dot{x} \\ \dot{\psi} \end{pmatrix} = \begin{pmatrix} A + BC & GR^{-1}G^* \\ -\lambda W(t) & -(A + BC)^* \end{pmatrix} \begin{pmatrix} x \\ \psi \end{pmatrix}, \tag{13}$$

$$x(t_0) = 0, \quad \psi(T) = 0, \tag{14}$$

for some λ such that $\int_{t_0}^T x^* W x \, dt > 0$. Also let $v \triangleq R^{-1}G^*\psi$. Then

$$\frac{\int_{t_0}^T v^*(t)R(t)v(t)\,dt}{\int_{t_0}^T x^*(t)W(t)x(t)\,dt} = \lambda. \tag{15}$$

Proof. We have

$$\int_{t_0}^T v^* R v \, dt = \int_{t_0}^T \psi^* G R^{-1} G^* \psi \, dt$$

$$= \int_{t_0}^T \psi^* \{\dot{x} - (A + BC)x\} \, dt.$$

Integrating the above expression by parts and using (14),

$$\int_{t_0}^T v^* R v \, dt = \lambda \int_{t_0}^T x^* W x \, dt. \tag{16}$$

$\square$

Since the optimal $v(t)$ satisfies (13) and (14), we deduce from Theorem 2.2 that the minimum value of (5) is the same as the minimum positive λ such that (13) and (14) have a solution with $\int_{t_0}^T x^* W x \, dt > 0$.

3. A NECESSARY CONDITION FOR THE MAXIMIZATION OF λ

We assume that there is a $C_0(t)$ which maximizes λ and stabilizes the system. Also assume that $C_0(t)$ has neighborhood in which the system maintains stability. To derive the necessary conditions of this section, we follow the procedure of Refs. [12] and [13]. Now the necessary conditions satisfied by $C_0(t)$ can be stated as follows.

THEOREM 3.1. *Consider the boundary value problem*

$$\dot{x}_0 = (A(t) + B(t)C(t))x_0 + G(t)R^{-1}(t)G^*(t)\psi_0,$$
$$\dot{\psi}_0 = -(A(t) + B(t)C(t))^*\psi_0 - \lambda W(t)x_0(t),$$
(17)

with

$$x_0(t_0) = 0, \quad \psi_0(T) = 0.$$
(18)

Let $C_0(t)$ maximize the minimum positive value of λ for which (17) and (18) have a solution with $\int_{t_0}^T x_0^* W x_0\, dt > 0$. Denote the value of λ corresponding to $C_0(t)$ by λ_0. Let $\delta C(t)$ denote an elemental perturbation of $C_0(t)$. Then $C_0(t)$ satisfies

$$\int_{t_0}^T (\psi_0^*(t)B(t) + \lambda_0 x_0^*(t)C_0^*(t)Q_2(t))\delta C(t)\, x_0(t)\, dt = 0.$$
(19)

Proof. Let the perturbations in $x_0(t)$, $\psi_0(t)$, and λ_0 corresponding to $\delta C(t)$ be denoted by $x_1(t)$, $\psi_1(t)$, and μ respectively. In the following equations, we omit to show the explicit dependence of the coefficient matrices on t for simplicity of notation. Let $\bar{G} = GR^{-1}G^*$. We have the following set of equations.

$$\dot{x}_0 = (A + BC_0)x_0 + \bar{G}\psi_0,$$
(20)

$$\dot{\psi}_0 = -(A + BC_0)^*\psi_0 - \lambda_0 W x_0,$$
(21)

$$x_0(t_0) = 0, \quad \psi_0(T) = 0,$$
(22)

$$\dot{x}_1 = (A + BC_0)x_1 + B\delta C\, x_0 + \bar{G}\psi_1,$$
(23)

$$\dot{\psi}_1 = -(A + BC_0)^*\psi_1 - (B\delta C)^*\psi_0 - \mu W x_0$$
$$\qquad\qquad - \lambda_0 W x_1 - \lambda_0(\delta C^* Q_2 C_0 + C_0^* Q_2 \delta C)x_0,$$
(24)

$$x_1(t_0) = 0, \quad \psi_1(T) = 0.$$
(25)

From (24) we get

$$\int_{t_0}^{T} x_0^* \dot{\psi}_1 \, dt = - \int_{t_0}^{T} x_0^* (A + BC_0)^* \psi_1 \, dt - \int_{t_0}^{T} x_0^* (B\delta C)^* \psi_0 \, dt$$

$$- \mu \int_{t_0}^{T} x_0^* W x_0 \, dt - \lambda_0 \int_{t_0}^{T} x_0^* W x_1 \, dt$$

$$- \lambda_0 \int_{t_0}^{T} x_0^* (\delta C^* Q_2 C_0 + C_0^* Q_2 \delta C) x_0 \, dt \qquad (26)$$

Integrating the left side of (26) by parts and making use of (20), we get

$$\int_{t_0}^{T} \psi_0^* \bar{G} \psi_1 \, dt = \int_{t_0}^{T} x_0^* (B\delta C)^* \psi_0 \, dt + \mu \int_{t_0}^{T} x_0^* W x_0 \, dt$$

$$+ \lambda_0 \int_{t_0}^{T} x_0^* W x_1 \, dt + \lambda_0 \int_{t_0}^{T} x_0^* (\delta C^* Q_2 C_0 + C_0^* Q_2 \delta C) x_0 \, dt. \qquad (27)$$

By equation (21), the third integral on the right side of (27) is written as

$$\lambda_0 \int_{t_0}^{T} x_0^* W x_1 \, dt = - \int_{t_0}^{T} \left(\dot{\psi} + (A + BC_0)^* \psi_0 \right)^* x_1 \, dt. \qquad (28)$$

An integration by parts and equation (23) yield

$$\lambda_0 \int_{t_0}^{T} x_0^* W x_1 \, dt = \int_{t_0}^{T} [\psi_0^* B\delta C \, x_0 + \psi_0^* \bar{G} \psi_1] \, dt. \qquad (29)$$

Since λ_0 is maximal, $\mu = 0$ in (27). Substituting (29) in (27) and simplifying, we get

$$\int_{t_0}^{T} (\psi_0^* B + \lambda_0 x_0^* C_0^* Q_2) \delta C \, x_0 \, dt = 0. \qquad (30)$$

$\square$

In the case of time-invariant systems with all the associated matrices being constant, the above equation implies that each component of x_0 is orthogonal to $B^* \psi_0 + \lambda_0 Q_2 C_0 x_0$ on $[t_0, T]$.

Since nonlinear programming algorithms are used to get C_0 and several nonlinear programming algorithms make use of gradients, it is useful to derive an expression for the variation of λ as a function of the variation in $C(t)$. For this, suppose that λ is the minimum value of (5) corresponding to $C(t)$ and μ is the perturbation in λ owing to an elemental perturbation $\delta C(t)$ of $C(t)$. Following the same procedure as in the proof of Theorem 3.1, we have

$$\mu = \frac{-2\int_{t_0}^{T}(\psi^* B + \lambda x^* C^* Q_2)\delta C\, x\, dt}{\int_{t_0}^{T} x^* W x\, dt}. \tag{31}$$

4. PROBLEM FORMULATION FOR AN
OBSERVER-BASED CONTROLLER

The system equations are given by

$$\dot{x} = A(t)x + B(t)u + G(t)v, \quad x(t_0) = 0, \tag{32}$$

$$y = C_2(t)x, \tag{33}$$

$$\dot{\hat{x}} = A(t)\hat{x} + B(t)u + L(t)[C_2(t)x - C_2(t)\hat{x}], \quad \hat{x}(t_0) = 0, \tag{34}$$

$$u = C(t)\hat{x}, \tag{35}$$

where $y(t)$ denotes the output vector and $\hat{x}$, the observer state. Assume that the observer gain $L(t)$ is given. For a given $C(t)$, let λ denote the minimum value of the performance index

$$J(v) = \frac{\int_{t_0}^{T} v^*(t)R(t)v(t)\, dt}{\int_{t_0}^{T} \{x^*(t)Q_1(t)x(t) + u^*(t)Q_2(t)u(t)\}\, dt}. \tag{36}$$

For maximal disturbance rejection, we need to choose $C(t)$ such that λ is maximized.

The above equations can be written as

$$\begin{pmatrix} \dot{x} \\ \dot{\hat{x}} \end{pmatrix} = \begin{pmatrix} A & BC \\ LC_2 & A + BC - LC_2 \end{pmatrix} \begin{pmatrix} x \\ \hat{x} \end{pmatrix} + \begin{pmatrix} G(t) \\ 0 \end{pmatrix} v, \tag{37}$$

$$x(t_0) = \hat{x}(t_0) = 0, \tag{38}$$

with the performance index being

$$J(v) = \frac{\int_{t_0}^{T} v^* R v \, dt}{\int_{t_0}^{T} (x^* \quad \hat{x}^*) \begin{pmatrix} Q_1 & 0 \\ 0 & C^* Q_2 C \end{pmatrix} \begin{pmatrix} x \\ \hat{x} \end{pmatrix} dt}. \tag{39}$$

5. Necessary Conditions for a Fixed $C(t)$
in the Case of an Observer-Based Controller

For fixed $C(t)$, if $v(t)$ minimizes (39), then it also minimizes the alternate cost functional

$$J_1(v) \triangleq \frac{1}{2} \int_{t_0}^{T} \{ v^* R v - \lambda (x^* Q_1 x + \hat{x}^* C^* Q_2 C \hat{x}) \} \, dt \tag{40}$$

where λ is the minimum value of (39). The necessary conditions for optimal $v(t)$ can be stated as follows.

THEOREM 5.1. *Consider the system given by* (32)-(35) *for fixed* $C(t)$ *with the cost functional given by* (36). *If* $v(t)$ *minimizes* (36), *then there exist adjoint vectors* $\psi(t)$ *and* $\hat{\psi}(t)$, *not both identically zero, such that*

$$\frac{d\psi}{dt} = -A^* \psi - (LC_2)^* \hat{\psi} - \lambda Q_1 x, \tag{41}$$

$$\frac{d\hat{\psi}}{dt} = -(BC)^* \psi - (A + BC - LC_2)^* \hat{\psi} - \lambda C^* Q_2 C \hat{x}, \tag{42}$$

$$\psi(T) = \hat{\psi}(T) = 0, \tag{43}$$

$$v(t) = R^{-1}(t) G^*(t) \psi(t), \tag{44}$$

where λ *is the minimum value of* (36).

Proof. The Hamiltonian is given by

$$H(\psi, \hat{\psi}, x, \hat{x}, v) = \psi^* (Ax + BC\hat{x} + Gv) + \hat{\psi}^* \{ (A + BC - LC_2)\hat{x} + LC_2 x \}$$
$$- \frac{1}{2} \{ v^* R v - \lambda (x^* Q_1 x + \hat{x}^* C^* Q_2 C \hat{x}) \} \tag{45}$$

where ψ and $\hat{\psi}$ are the adjoint vectors.

The worst disturbance is obtained by setting $\partial H/\partial v = 0$ and is given by

$$v(t) = R^{-1}(t)G^*(t)\psi(t). \tag{46}$$

The adjoint vectors ψ and $\hat{\psi}$ satisfy

$$\frac{d\psi}{dt} = -\frac{\partial H}{\partial x} = -A^*\psi - (LC_2)^*\hat{\psi} - \lambda Q_1 x, \tag{47}$$

$$\frac{d\hat{\psi}}{dt} = -\frac{\partial H}{\partial \hat{x}} = -(BC)^*\psi - (A + BC - LC_2)^*\hat{\psi} - \lambda C^*Q_2 C\hat{x}, \tag{48}$$

with

$$\psi(T) = \hat{\psi}(T) = 0. \tag{49}$$

$\square$

Thus we have a two-point boundary value problem given by (37), (38), and (41)-(44). Using the same technique as in Section 2, it can be shown that the minimum value of (36) is the least positive λ for which the boundary value problem has a solution with $\int_{t_0}^T \{x^*Q_1 x + \hat{x}^*C^*Q_2 C\hat{x}\}\, dt > 0$.

6. MAXIMIZATION OF λ IN THE

CASE OF AN OBSERVER-BASED CONTROLLER

In this section, for simplicity of notation, we omit denoting the matrices and vectors as functions of t. Assume that $C(t)$ maximizes λ and there is a neighborhood of $C(t)$ in which the system given by (37) maintains stability. We now derive a necessary condition that is satisfied by $C(t)$.

Let $\delta C(t)$ denote an elemental perturbation in $C(t)$. Let the corresponding perturbations in $x, \hat{x}, \psi, \hat{\psi}$, and λ be denoted by $x_1, \hat{x}_1, \psi_1, \hat{\psi}_1$, and μ respectively. Let $\bar{G} = GR^{-1}G^*$.

The equations and the boundary conditions of the variables and their perturbations are given below:

$$\dot{x} = Ax + BC\hat{x} + \bar{G}\psi, \tag{50}$$

$$\dot{\hat{x}} = LC_2 x + (A + BC - LC_2)\hat{x}, \tag{51}$$

$$\dot{\psi} = -A^*\psi - (LC_2)^*\hat{\psi} - \lambda Q_1 x, \tag{52}$$

$$\dot{\hat{\psi}} = -(BC)^*\psi - (A + BC - LC_2)^*\hat{\psi} - \lambda C^* Q_2 C\hat{x}, \tag{53}$$

$$x(t_0) = \hat{x}(t_0) = \psi(T) = \hat{\psi}(T) = 0, \tag{54}$$

$$\dot{x}_1 = Ax_1 + BC\hat{x}_1 + B\delta C\,\hat{x} + \bar{G}\psi_1, \tag{55}$$

$$\dot{\hat{x}}_1 = LC_2 x_1 + (A + BC - LC_2)\hat{x}_1 + B\delta C\hat{x}, \tag{56}$$

$$\dot{\psi}_1 = -A^*\psi_1 - (LC_2)^*\hat{\psi}_1 - \lambda Q_1 x_1 - \mu Q_1 x, \tag{57}$$

$$\dot{\hat{\psi}}_1 = -(BC)^*\psi_1 - (B\delta C)^*\psi - (A + BC - LC_2)^*\hat{\psi}_1$$
$$\qquad - (B\delta C)^*\hat{\psi} - \lambda C^* Q_2 C\hat{x}_1 - \mu C^* Q_2 C\hat{x} - \lambda(\delta C^* Q_2 C + C^* Q_2 \delta C)\hat{x}, \tag{58}$$

$$x_1(t_0) = \hat{x}_1(t_0) = \psi_1(T) = \hat{\psi}_1(T) = 0. \tag{59}$$

The necessary condition satisfied by $C(t)$ is given in the following theorem.

THEOREM 6.1. *Consider the boundary value problem given by (50)-(54). Let $C(t)$ maximize the minimum positive value for which (50)-(54) has a solution with $\int_{t_0}^T \{x^* Q_1 x + \hat{x}^* C^* Q_2 C\hat{x}\}\, dt > 0$. Denote the maximum of the minimum positive value by λ. Let $\delta C(t)$ denote an arbitrary elemental perturbation in $C(t)$. Then $C(t)$ satisfies*

$$\int_{t_0}^T \{(\psi + \hat{\psi})^* B + \lambda \hat{x}^* C^* Q_2\}\delta C\,\hat{x}\, dt = 0. \tag{60}$$

Proof. To derive the necessary condition satisfied by $C(t)$, we proceed as follows. From (57), we get

$$\int_{t_0}^T x^* \dot{\psi}_1\, dt = -\int_{t_0}^T (x^* A^* \psi_1 + x^*(LC_2)^* \hat{\psi}_1 + \lambda x^* Q_1 x_1 + \mu x^* Q_1 x)\, dt. \tag{61}$$

By integrating the left side of (61) by parts, we get

$$\int_{t_0}^{T} x^* \dot{\psi}_1 \, dt = - \int_{t_0}^{T} (x^* A^* \psi_1 + \hat{x}^*(BC)^* \psi_1 + \psi^* \bar{G} \psi_1) \, dt. \tag{62}$$

Also, by (58)

$$\int_{t_0}^{T} \hat{x}^* \dot{\psi}_1 \, dt = - \int_{t_0}^{T} \{ \hat{x}^*(BC)^* \psi_1 + \hat{x}^*(B\delta C)^* \psi + \hat{x}^*(A + BC - LC_2)^* \hat{\psi}_1$$

$$+ \hat{x}^*(B\delta C)^* \hat{\psi} + \lambda \hat{x}^* C^* Q_2 C \hat{x}_1 + \mu \hat{x}^* C^* Q_2 C \hat{x}$$

$$+ \lambda \hat{x}^*(\delta C^* Q_2 C + C^* Q_2 \delta C) \hat{x} \} \, dt. \tag{63}$$

By an integration by parts, the left side of (63) is

$$\int_{t_0}^{T} \hat{x}^* \dot{\hat{\psi}}_1 \, dt = - \int_{t_0}^{T} (\hat{x}^*(A + BC - LC_2)^* \hat{\psi}_1 + x^*(LC_2)^* \hat{\psi}_1) \, dt. \tag{64}$$

Adding the right sides of (62) and (64), and equating it to the sum of the right sides of (61) and (63), and simplifying

$$\int_{t_0}^{T} \psi^* \bar{G} \psi_1 \, dt = \lambda \int_{t_0}^{T} x^* Q_1 x_1 \, dt + \int_{t_0}^{T} \hat{x}^*(B\delta C)^*(\psi + \hat{\psi}) \, dt$$

$$+ \lambda \int_{t_0}^{T} \hat{x}^* C^* Q_2 C \hat{x}_1 \, dt + \lambda \int_{t_0}^{T} \hat{x}^*(\delta C^* Q_2 C + C^* Q_2 \delta C) \hat{x} \, dt$$

$$+ \mu \int_{t_0}^{T} (x^* Q_1 x + \hat{x}^* C^* Q_2 C \hat{x}) \, dt. \tag{65}$$

From (52) and (53),

$$\lambda Q_1 x = -(\dot{\psi} + A^* \psi + (LC_2)^* \hat{\psi}), \tag{66}$$

and

$$\lambda C^* Q_2 C \hat{x} = -(\dot{\hat{\psi}} + (BC)^* \psi + (A + BC - LC_2)^* \hat{\psi}). \tag{67}$$

Using (66) and (67) and integrating by parts, we get

$$\lambda \int_{t_0}^{T} x^* Q_1 x_1 \, dt = \int_{t_0}^{T} \psi^*(BC\hat{x}_1 + B\delta C \hat{x} + \bar{G}\psi_1) \, dt - \int_{t_0}^{T} \hat{\psi}^*(LC_2)x_1 \, dt \qquad (68)$$

and

$$\lambda \int_{t_0}^{T} \hat{x}^* C^* Q_2 C \hat{x}_1 \, dt = \int_{t_0}^{T} \hat{\psi}^*(LC_2 x_1 + B\delta C \hat{x}) \, dt - \int_{t_0}^{T} \psi^*(BC)\hat{x}_1 \, dt. \qquad (69)$$

Incorporating (68) and (69) in (65) and simplifying, we get (letting $\mu = 0$)

$$\int_{t_0}^{T} \{(\psi + \hat{\psi})^* B + \lambda \hat{x}^* C^* Q_2\} \delta C \, \hat{x} \, dt = 0 \qquad (70)$$

for arbitrary $\delta C(t)$.   □

If $C(t)$ is nonoptimal, an expression for the variation μ in λ can be derived in terms of variation in $C(t)$. Following the same reasoning as in Theorem 6.1, we have

$$\mu = \frac{-2 \int_{t_0}^{T} \{(\psi + \hat{\psi})^* B + \lambda \hat{x}^* C^* Q_2\} \delta C \, \hat{x} \, dt}{\int_{t_0}^{T} (x^* Q_1 x + \hat{x}^* C^* Q_2 C \hat{x}) \, dt}. \qquad (71)$$

7. REDUCED ORDER OBSERVER CASE

The equations are given by

$$\dot{x} = A(t)x + B(t)u + G(t)v, \quad x(t_0) = 0, \qquad (72)$$

$$y = C_2(t)x, \qquad (73)$$

$$\dot{x}_r = F(t)x_r + M(t)u + N(t)y, \quad x_r(t_0) = 0, \qquad (74)$$

$$u = C(t)x_r + D(t)y, \qquad (75)$$

where x_r is the state vector corresponding to the reduced state observer. For given $C(t)$ and $D(t)$, let λ denote the minimum value of

$$J(v) = \frac{\int_{t_0}^{T} v^*(t)R(t)v(t) \, dt}{\int_{t_0}^{T} \{x^*(t)Q_1(t)x(t) + u^*(t)Q_2(t)u(t)\} \, dt}. \qquad (76)$$

The problem is to choose $C(t)$ and $D(t)$ such that λ is maximized.

Let $z = (x^*, x_r^*)^*$. Then equations(72)-(76) can be written as

$$\dot{z} = \tilde{A}(t)z + \tilde{G}(t)v, \quad z(t_0) = 0, \tag{77}$$

where

$$\tilde{A} = \begin{pmatrix} A + BDC_2 & BC \\ MDC_2 + NC_2 & F + MC \end{pmatrix}, \quad \tilde{G} = \begin{pmatrix} G \\ 0 \end{pmatrix}, \tag{78}$$

with the cost functional being

$$\frac{\int_{t_0}^T v^* R v \, dt}{\int_{t_0}^T z^* W z \, dt}, \tag{79}$$

where

$$W = \begin{pmatrix} Q_1 + C_2^* D^* Q_2 D C_2 & C_2^* D^* Q_2 C \\ C^* Q_2 D C_2 & C^* Q_2 C \end{pmatrix}. \tag{80}$$

Application of the maximum principle yields $v(t)$ which minimizes (79) as

$$v(t) = R^{-1} \tilde{G}^* \psi, \tag{81}$$

where ψ satisfies

$$\dot{\psi} = -\tilde{A}^* \psi - \lambda W z, \quad \psi(T) = 0. \tag{82}$$

Again it can be shown that the minimum value of (79) is the least value of λ such that (77), (81), and (82) have a solution with $\int_{t_0}^T z^* W z \, dt > 0$.

Suppose $C(t)$ and $D(t)$ maximize λ. Let δC and δD denote elemental perturbations in C and D respectively, with the corresponding perturbations in z and ψ being z_1 and ψ_1. Let $\bar{G} = \tilde{G} R^{-1} \tilde{G}^*$. Also denote the perturbations in $\tilde{A}, W$, and λ by $\delta\tilde{A}, \delta W$, and μ respectively. Thus, we have the following set of equations:

$$\dot{z} = \tilde{A} z + \bar{G}\psi, \tag{83}$$

$$\dot{\psi} = -\tilde{A}^*\psi - \lambda W z, \tag{84}$$

$$z(t_0) = \psi(T) = 0, \tag{85}$$

$$\dot{z}_1 = \tilde{A} z_1 + \bar{G}\psi_1 + \delta\tilde{A} z, \tag{86}$$

$$\dot{\psi}_1 = -\tilde{A}^*\psi_1 - \lambda W z_1 - \delta\tilde{A}^*\psi - \mu W z - \lambda\delta W z, \tag{87}$$

$$z_1(t_0) = \psi_1(T) = 0. \tag{88}$$

The necessary conditions satisfied by $C(t)$ and $D(t)$ are stated in the following theorem.

THEOREM 7.1. *Consider the boundary value problem given by (83)-(85). Assume that $C(t)$ and $D(t)$ maximize the minimum positive value for which (83)-(85) has a solution with $\int_{t_0}^{T} z^* W z\, dt > 0$. Denote the maximum of the minimum positive value by λ. Let $\delta C(t)$ and $\delta D(t)$ denote elemental perturbations in $C(t)$ and $D(t)$ respectively. Let $\bar{\psi}$ and ψ_r denote the components of ψ corresponding to x and x_r respectively. Then $C(t)$ and $D(t)$ satisfy*

$$\int_{t_0}^{T} \{\bar{\psi}^* B + \psi_r^* M + \lambda x^* C_2^* D^* Q_2 + \lambda x_r^* C^* Q_2\}\delta D\, y\, dt = 0, \tag{89}$$

$$\int_{t_0}^{T} \{\bar{\psi}^* B + \psi_r^* M + \lambda x^* C_2^* D^* Q_2 + \lambda x_r^* C^* Q_2\}\delta C\, x_r\, dt = 0. \tag{90}$$

Proof. From (87),

$$\int_{t_0}^{T} z^* \dot{\psi}_1\, dt = -\int_{t_0}^{T} \{z^* \tilde{A}^*\psi_1 + \lambda z^* W z_1$$

$$+ z^* \delta\tilde{A}^*\psi + \mu z^* W x + \lambda z^* \delta W\, z\}\, dt. \tag{91}$$

Also by an integration by parts,

$$\int_{t_0}^{T} z^* \dot{\psi}_1\, dt = -\int_{t_0}^{T} \{z^* \tilde{A}^*\psi_1 + \psi^* \bar{G}\psi_1\}\, dt. \tag{92}$$

From (91) and (92),

$$\int_{t_0}^{T} \psi^* \bar{G} \psi_1 \, dt = \int_{t_0}^{T} \{\lambda z^* W z_1 + z^* \delta \tilde{A}^* \psi + \mu z^* W z + \lambda z^* \delta W \, z\} \, dt. \tag{93}$$

From (84),

$$\lambda W z = -(\dot{\psi} + \tilde{A}^* \psi). \tag{94}$$

Note that $\mu = 0$. Substituting (94) in (93) and integrating by parts,

$$\int_{t_0}^{T} \{\psi^* \delta \tilde{A} \, z + z^* \delta \tilde{A}^* \psi + \lambda z^* \delta W \, z\} \, dt = 0. \tag{95}$$

Equations (89) and (90) follow by substituting the expressions for $\delta \tilde{A}$ and δW in (95).   □

We remark that Theorem 6.1 can be proved utilizing a similar line of reasoning as in the proof of Theorem 7.1. Also, if $C(t)$ is nonoptimal, an expression for the variation μ in λ can be derived in terms of variations $\delta \tilde{A}$ and δW, which are a result of variations in $C(t)$ and $D(t)$. Following similar reasoning as in the proof of Theorem 7.1, we have

$$\mu = \frac{-\int_{t_0}^{T} \{\psi^* \delta \tilde{A} \, z + z^* \delta \tilde{A}^* \psi + \lambda z^* \delta W \, z\} \, dt}{\int_{t_0}^{T} z^* W z \, dt}. \tag{96}$$

8. SUFFICIENCY THEORY

Continuing the material of Section 7 with the same notation, denote the variations of second order in the variables z, ψ, and λ by z_2, ψ_2, and η respectively. The equations satisfied by z_2 and ψ_2 are

$$\dot{z}_2 = \tilde{A} z_2 + \bar{G} \psi_2 + 2 \delta \tilde{A} \, z_1, \tag{97}$$

$$\dot{\psi}_2 = -\tilde{A}^* \psi_2 - \lambda W z_2 - 2 \delta \tilde{A}^* \psi_1 - 2 \lambda \delta W \, z_1 - \eta W z - \lambda \delta^2 W \, z, \tag{98}$$

$$z_2(t_0) = \psi_2(T) = 0. \tag{99}$$

From (98),

$$\int_{t_0}^{T} z^* \dot{\psi}_2 \, dt = - \int_{t_0}^{T} \{ z^* \tilde{A}^* \psi_2 + \lambda z^* W z_2 + 2 z^* \delta \tilde{A}^* \psi_1$$

$$+ 2 \lambda z^* \delta W z_1 + \eta z^* W z + \lambda z^* \delta^2 W z \} \, dt. \qquad (100)$$

Also, by an integration by parts

$$\int_{t_0}^{T} z^* \dot{\psi}_2 \, dt = - \int_{t_0}^{T} \{ z^* \tilde{A}^* \psi_2 + \psi^* \bar{G} \psi_2 \} \, dt. \qquad (101)$$

However, from (97)

$$\bar{G} \psi_2 = \dot{z}_2 - \tilde{A} z_2 - 2 \delta \tilde{A} z_1. \qquad (102)$$

Substituting (102) in (101) and integrating by parts, we get

$$\int_{t_0}^{T} z^* \dot{\psi}_2 \, dt = - \int_{t_0}^{T} \{ z^* \tilde{A}^* \psi_2 + \lambda z^* W z_2 - 2 \psi^* \delta \tilde{A} z_1 \} \, dt. \qquad (103)$$

From (100) and (103),

$$\eta \int_{t_0}^{T} z^* W z \, dt = -2 \int_{t_0}^{T} \{ \psi^* \delta \tilde{A} z_1 + z^* \delta \tilde{A}^* \psi_1 + \lambda z^* \delta W z_1 \} - \lambda \int_{t_0}^{T} z^* \delta^2 W z \, dt. \qquad (104)$$

Note that the right side of (104) is independent of the second variations in z and ψ. A sufficient condition for the local maximization of λ is that $\eta \leq 0$. Finding conditions that ensure the nonpositivity of the right side of (104) is an open problem.

9. AN EXAMPLE

The necessary conditions are useful in verifying the optimality of the control matrix $C(t)$ which gives maximum value to λ. For time-invariant systems, optimization routines are needed to determine C. We now present an example.

Consider the scalar system

$$\dot{x} = -x + u + v, \quad x(0) = 0, \quad u = cx. \tag{105}$$

The objective is to choose c which maintains the stability of the system and maximizes the minimum of

$$\frac{\int_0^1 v^2(t)\, dt}{\int_0^1 (x^2 + u^2)\, dt} = \frac{\int_0^1 v^2(t)\, dt}{(1 + c^2)\int_0^1 x^2(t)\, dt}. \tag{106}$$

From equations (11) and (12), we get

$$\begin{pmatrix} \dot{x}_0 \\ \dot{\psi}_0 \end{pmatrix} = \begin{pmatrix} c - 1 & 1 \\ -\lambda(1 + c^2) & -(c - 1) \end{pmatrix} \begin{pmatrix} x_0 \\ \psi_0 \end{pmatrix}, \tag{107}$$

with

$$x_0(0) = 0, \quad \psi_0(1) = 0. \tag{108}$$

Let λ be the least positive value such that (107) and (108) have a nonzero solution. We need to find c which maximizes λ. Using an optimization routine to get the optimum c, we have $c_0 = -0.34$ with $\lambda_0 = 5.68371$.

In this case (30) becomes

$$\int_0^1 \psi_0 x_0\, dt = -\lambda_0 c_0 \int_0^1 x_0^2\, dt. \tag{109}$$

Equation (109) can be easily verified.

10. CONCLUSIONS

In this chapter we derived necessary conditions for optimal disturbance rejection. The conditions are useful in synthesizing a controller which maximizes the disturbance rejection capacity of a system. Necessary conditions are also derived in the case of an observer-based controller. The design procedure of this chapter is useful in the synthesis of a controller for which the effect of gust on the airplane performance during landing is as small as possible. Also, in high speed low altitude terrain following, the methodology can be applied to attenuate the effect of disturbances in an optimal manner.

References

[1] J. DOYLE, K. GLOVER, P. KHARGONEKAR AND B. FRANCIS, State-space solutions to standard H_2 and H_∞ control problems, *IEEE Trans. Automat. Contr.* **34**, 1989, pp. 831-847.

[2] K. GLOVER AND J. C. DOYLE, State space formulae for all stabilizing controllers that satisfy an H_∞-norm bound and relations to risk sensitivity, *Systems and Control Letters* **11**, 1988, pp. 167-172.

[3] H. KIMURA AND R. KAWATANI, "Synthesis of H^∞ controllers based on conjugation," Proc. 27th IEEE Conference on Decision and Control, 1988, pp. 7-13.

[4] D. S. BERNSTEIN AND W. M. HADDAD, LQG control with an H_∞ performance bound : A Riccati equation approach, *IEEE Trans. Automat. Contr.* **34**, 1989, pp. 293-305.

[5] P. P. KHARGONEKAR, I. R. PETERSEN AND M. A. ROTEA, H_∞-optimal control with state-feedback, *IEEE Trans. Automat. Contr.* **33**, 1988, pp. 786-788.

[6] P. P. KHARGONEKAR, I. R. PETERSEN AND K. ZHOU, Robust stabilization and H_∞-optimal control, 1987.

[7] G. TADMOR, H_∞ in the time domain : The standard four block problem, *Mathematics of Control, Signals, and Systems*, to be published.

[8] M. B. SUBRAHMANYAM, On integral inequalities associated with a linear operator equation, *Proc. Amer. Math. Soc.* **92**, 1984, pp. 342-346.

[9] _____ , On applications of control theory to integral inequalities: II, *SIAM J. Contr. Optimiz.* **19**, 1981, pp. 479-489.

[10] _____ , Necessary conditions for minimum in problems with nonstandard cost functionals, *J. Math. Anal. Appl.* **60**, 1977, pp. 601-616.

[11] E. B. LEE AND L. MARKUS, "Foundations of Optimal Control Theory," Wiley, New York, 1967.

[12] A. S. BRATUS AND A. P. SEIRANYAN, Sufficient conditions for an extremum in eigen-value optimization problems, *PMM U.S.S.R. (Journal of Applied Mathematics and Mechanics, English translation)* **48**, 1984, pp. 466-474.

[13] I. TADJBAKHSH AND J. B. KELLER, Strongest columns and isoperimetric inequalities for eigenvalues, *ASME J. Appl. Mech.* **29**, 1962, pp. 159-164.

CHAPTER 5

Synthesis of Finite-Interval H_∞ Controllers
by State Space Methods

ABSTRACT

In this chapter a state space formulation of the H_∞ optimal control problem is given. Assuming a finite interval of control, the problem of synthesizing a finite-interval H_∞ controller is converted into an optimization problem in which a parameter occurring in a boundary value problem needs to be maximized. An optimality condition for the maximization of this parameter is given. The proposed method makes use of the observer-based parametrization of all stabilizing controllers. An example is worked out.

1. INTRODUCTION

The H_∞ optimal control theory has been pioneered by Zames [1] and important contributions have been made by Francis and Doyle [2,3]. Recent work [4] indicates that the theory has important applications in the design of flight control systems.

In this chapter a variant of the H_∞ problem is considered in terms of state space formulation. Optimization routines are needed for the synthesis of the final controller. The formulation is based on considering optimal control problems with finite terminal time in which the cost is a quotient of two definite integrals. The mathematical theory behind the method is given in Chapters 1-3.

Other authors have considered the H_∞ problem from different points of view. In [5] a parametrization of all stabilizing controllers that achieve a specified H_∞ norm bound is given in a specialized case. The computation of the controller involves the solution of two Riccati equations. This result has been extended to the general case in [6]. In [7] the H_∞ problem is solved by introducing a generalized algebraic operation called conjugation. The approach again yields two Riccati equations whose solution leads to the synthesis of a controller. In [8] a certain LQG problem with a side constraint on the H_∞-norm of the closed loop transfer function is solved. In this approach it is necessary to solve three coupled Riccati equations. In special cases these three equations can be reduced to two Riccati equations.

Our approach is similar to that in Chapter 3 and it results in a two-point boundary value problem. The approach has the advantage of being applicable to time-varying systems with observer-based controllers and dynamic controllers. Ref. 9 contains one such application in which the objective is to maximize the disturbance rejection capacity of a time-varying linear system. The material of [9] is essentially presented in Chapter 3. Also, given a controller it is important to know the performance measure of the controller. For the general time-varying system with a given controller, the parameter λ of Section 3 gives a measure of the performance of the controller.

Our time-domain approach has several advantages even in the case of time-invariant systems. First of all, it provides an alternate new approach to the computation of finite-interval H_∞ controllers. The H_∞ algorithms usually cannot handle time domain specifications. In our optimization algorithm it is possible to include time domain constraints. Also time domain approach is convenient for handling parameter uncertainties. In Chapters 3 and 6, we address the important robust performance problem, viz., how to achieve maximum performance and required robustness under parameter variations.

2. STATE SPACE FORMULATION OF THE H_∞ PROBLEM

The standard H_∞ problem can be stated with reference to Fig. 1 (p. 87). In Fig. 1 w, u, z, and y denote the exogenous input (command signals, disturbances, sensor noises etc.), the control input, the output to be controlled, and the measured output, respectively. The plant $G(s)$ and the controller $K(s)$ are assumed to be real-rational and proper. Partition G as

$$G = \begin{pmatrix} G_{11} & G_{12} \\ G_{21} & G_{22} \end{pmatrix}.$$
(1)

The equations corresponding to Fig. 1 are

$$z = G_{11}w + G_{12}u, \quad y = G_{21}w + G_{22}u, \quad u = Ky.$$
(2)

The standard H_∞ problem is to find a real-rational proper K which minimizes the H_∞ norm of the transfer matrix from w to z under the constraint that K stabilize G.

In terms of state space equations $G(s)$ is written as

$$\dot{x} = Ax + B_1 w + B_2 u$$

$$z = C_1 x + D_{11}w + D_{12}u$$
(3)

$$y = C_2 x + D_{21}w + D_{22}u.$$

Doyle [10] showed that every stabilization procedure can be realized as an observer-based controller by adding stable dynamics to the plant. The realization of the observer-based controller is shown in Fig. 2 (p. 87) where the stable dynamics added is represented by $Q(s)$, with $Q(s)$ proper and $I - D_{22}Q(\infty)$ invertible. In Fig. 2, F and H are chosen such that $A + B_2 F$ and $A + HC_2$ are stable. Assume that $Q(s)$ is described by the minimal representation

$$\dot{q} = \tilde{A}q + \tilde{B}\tilde{y}, \quad u_2 = \tilde{C}q + \tilde{D}\tilde{y}.$$
(4)

Following the notation of [11], define the following quantities.

$$\beta_1 = -H - (B_2 + HD_{22})(I - \tilde{D}D_{22})^{-1}\tilde{D}$$

$$\beta_2 = \tilde{B} + \tilde{B}D_{22}(I - \tilde{D}D_{22})^{-1}\tilde{D}$$

$$\gamma_1 = F + (I - \tilde{D}D_{22})^{-1}\tilde{D}(C_2 + D_{22}F)$$

$$\gamma_2 = -(I - \tilde{D}D_{22})^{-1}\tilde{C}$$

$$\alpha_{11} = A + HC_2 + (B_2 + HD_{22})\gamma_1$$

$$= A + B_2F - \beta_1(C_2 + D_{22}F)$$

$$\alpha_{12} = (B_2 + HD_{22})\gamma_2$$

$$\alpha_{21} = -\beta_2(C_2 + D_{22}F)$$

$$\alpha_{22} = \tilde{A} - \tilde{B}D_{22}\gamma_2$$

$$\kappa = -(I - \tilde{D}D_{22})^{-1}\tilde{D}.$$

$$(5)$$

Then the closed loop system is given by

$$\begin{pmatrix} \dot{x} \\ \dot{\hat{x}} \\ \dot{q} \end{pmatrix} = \begin{pmatrix} A_{11} & A_{12} & A_{13} \\ A_{21} & A_{22} & A_{23} \\ A_{31} & A_{32} & A_{33} \end{pmatrix} \begin{pmatrix} x \\ \hat{x} \\ q \end{pmatrix} + \begin{pmatrix} B_1 \\ B_2 \\ B_3 \end{pmatrix} w, \tag{6}$$

$$y = (I - D_{22}\kappa)^{-1}[C_2x + D_{22}\gamma_1\hat{x} + D_{22}\gamma_2q + D_{21}w], \tag{7}$$

$$z = C_1x + D_{11}w + D_{12}(\gamma_1\hat{x} + \gamma_2q + \kappa y), \tag{8}$$

where

$$A_{11} = A + B_2\kappa(I - D_{22}\kappa)^{-1}C_2$$

$$A_{12} = B_2\gamma_1 + B_2\kappa(I - D_{22}\kappa)^{-1}D_{22}\gamma_1$$

$$A_{13} = B_2\gamma_2 + B_2\kappa(I - D_{22}\kappa)^{-1}D_{22}\gamma_2$$

$$A_{21} = \beta_1(I - D_{22}\kappa)^{-1}C_2$$

$$A_{22} = \alpha_{11} + \beta_1(I - D_{22}\kappa)^{-1}D_{22}\gamma_1$$

$$A_{23} = \alpha_{12} + \beta_1(I - D_{22}\kappa)^{-1}D_{22}\gamma_2 \tag{9}$$

$$A_{31} = \beta_2(I - D_{22}\kappa)^{-1}C_2$$

$$A_{32} = \alpha_{21} + \beta_2(I - D_{22}\kappa)^{-1}D_{22}\gamma_1$$

$$A_{33} = \alpha_{22} + \beta_2(I - D_{22}\kappa)^{-1}D_{22}\gamma_2$$

$$B_1 = B_1 - B_2\tilde{D}(I - D_{22}\kappa)^{-1}D_{21}$$

$$B_2 = -(H + B_2\tilde{D})D_{21}$$

$$B_3 = \tilde{B}D_{21}.$$

Consider equations (4)-(9). Now the H_∞ control problem is to find among all sets of matrices $\tilde{A}, \tilde{B}, \tilde{C}$, and $\tilde{D}$ which give a stable transfer matrix from $\tilde{y}$ to u_2 (see Fig. 2) one for which the H_∞-norm of the transfer matrix from w to z is minimized.

The above problem is equivalent to the following problem. Suppose $\tilde{A}$ is selected to be a stable matrix. For fixed $\tilde{A}, \tilde{B}, \tilde{C}$, and $\tilde{D}$, let

$$\lambda = \inf_{w} \frac{\int_0^\infty w^*(t)w(t)\, dt}{\int_0^\infty z^*(t)z(t)\, dt}, \tag{10}$$

where the superscript $*$ denotes matrix or vector transpose. Now find the values of $\tilde{A}, \tilde{B}, \tilde{C}$, and $\tilde{D}$ which make λ a maximum. The initial conditions for the variables $x, \hat{x}$, and q are of course zero.

It is clear that the H_∞-norm of the transfer function from w to z is $1/\sqrt{\lambda}$ and the objective is to minimize the H_∞-norm by choosing a controller. However, since the basic theory for cost functionals of the form of a quotient of definite integrals is given in Chapters 1-3, we follow the same procedure as in these chapters and consider the equivalent problem of maximizing λ.

The input $w(t)$ considered in the above problem is an element of $L_2(0, \infty)$. However, in many physical systems, the control interval is finite. For example, in the case of an advanced fighter, most maneuvers are accomplished in the course of a few seconds. Thus, in the next section we consider an approximate H_∞ problem in the sense that the control interval will be finite. If the integration limit T in, say equation (13), approaches infinity, then $\sqrt{\lambda}$ is the inverse of the H_∞-norm of the transfer matrix from w to z. For lack of a better term, we call this a finite-interval H_∞ problem. On the other hand, the problem will be more general in the sense that time-varying linear systems and a broader class of performance indices will be considered in Section 3.

To motivate the problem considered in the next section, let $\mathbf{x} = (x^*, \hat{x}^*, q^*)^*$. Equations (6)-(8) are written as

$$\dot{\mathbf{x}} = \mathbf{A}\mathbf{x} + \mathbf{B}\mathbf{w}, \quad \mathbf{x}(0) = 0, \quad \mathbf{w} = w, \tag{11}$$

$$\mathbf{z} = z = \mathbf{C}\mathbf{x} + \mathbf{D}\mathbf{w}, \tag{12}$$

where the matrices $\mathbf{A}, \mathbf{B}, \mathbf{C}$, and $\mathbf{D}$ depend on $\tilde{A}, \tilde{B}, \tilde{C}$, and $\tilde{D}$. Let the control interval be $[0, T]$. For fixed $\tilde{A}, \tilde{B}, \tilde{C}$, and $\tilde{D}$ with $\tilde{A}$ being stable, let

$$\lambda = \inf_{\mathbf{w}} \frac{\int_0^T \mathbf{w}^*(t)\mathbf{w}(t)\, dt}{\int_0^T \mathbf{z}^*(t)\mathbf{z}(t)\, dt}. \tag{13}$$

Using an optimization routine, find the matrices $\tilde{A}, \tilde{B}, \tilde{C}$, and $\tilde{D}$ for which λ is maximized.

3. OPTIMALITY CONDITIONS

In this section we develop conditions for determining λ in a general case which subsumes the problem considered at the end of Section 2. These conditions will be developed for time-varying systems. The system equations are given by

$$\dot{\mathbf{x}} = \mathbf{A}(t)\mathbf{x} + \mathbf{B}(t)\mathbf{w}, \quad \mathbf{x}(t_0) = 0. \tag{14}$$

The problem on hand is to select $\mathbf{w}$ which minimizes the performance index given by

$$J(\mathbf{w}) = \frac{\int_{t_0}^{T}\{\frac{1}{2}\mathbf{x}^*\mathbf{R}_1\mathbf{x} + \mathbf{x}^*\mathbf{R}_2\mathbf{w} + \frac{1}{2}\mathbf{w}^*\mathbf{R}_3\mathbf{w}\}\,dt}{\int_{t_0}^{T}\{\frac{1}{2}\mathbf{x}^*\mathbf{W}_1\mathbf{x} + \mathbf{x}^*\mathbf{W}_2\mathbf{w} + \frac{1}{2}\mathbf{w}^*\mathbf{W}_3\mathbf{w}\}\,dt}. \tag{15}$$

Note that the performance index given by (13) can be regarded as a special case of (15) since $\mathbf{z} = \mathbf{C}\mathbf{x} + \mathbf{D}\mathbf{w}$. To get the performance index of (13), set $\mathbf{R}_1 = \mathbf{R}_2 = 0, \mathbf{W}_1 = \mathbf{C}^*\mathbf{C}, \mathbf{W}_2 = \mathbf{C}^*\mathbf{D}$, and $\mathbf{W}_3 = \mathbf{D}^*\mathbf{D}$ in equation (15). In (15) we assume that the weighting matrices $\mathbf{R}_1, \mathbf{R}_3, \mathbf{W}_1$, and $\mathbf{W}_3$ are symmetric and the integrands of both the numerator and the denominator are nonnegative for each $\mathbf{w}(t)$. Further, we assume that there is some $\mathbf{w}(t)$ for which the denominator is positive. Let $\lambda = \inf_{\mathbf{w}} J(\mathbf{w})$. We also assume that $\mathbf{R}_3 - \lambda\mathbf{W}_3$ is nonsingular.

Cost functionals of the form of (15) have been the subject matter of this monograph. For the sake of completeness, we derive the necessary conditions satisfied by an optimal $\mathbf{w}(t)$.

Since the infimum of (15) is λ, we have

$$\int_{t_0}^{T}\{\frac{1}{2}\mathbf{x}^*\mathbf{R}_1\mathbf{x} + \mathbf{x}^*\mathbf{R}_2\mathbf{w} + \frac{1}{2}\mathbf{w}^*\mathbf{R}_3\mathbf{w}\}\,dt$$
$$-\lambda\int_{t_0}^{T}\{\frac{1}{2}\mathbf{x}^*\mathbf{W}_1\mathbf{x} + \mathbf{x}^*\mathbf{W}_2\mathbf{w} + \frac{1}{2}\mathbf{w}^*\mathbf{W}_3\mathbf{w}\}\,dt \geq 0 \tag{16}$$

for all $(\mathbf{w}, \mathbf{x})$ which satisfy (14). Thus, if $\mathbf{w}$ minimizes the cost functional in (15), it also minimizes the alternate cost functional

$$J_1(\mathbf{w}) = \int_{t_0}^{T} \{\frac{1}{2}\mathbf{x}^*(\mathbf{R}_1 - \lambda\mathbf{W}_1)\mathbf{x} + \mathbf{x}^*(\mathbf{R}_2 - \lambda\mathbf{W}_2)\mathbf{w} + \frac{1}{2}\mathbf{w}^*(\mathbf{R}_3 - \lambda\mathbf{W}_3)\mathbf{w}\} \, dt. \quad (17)$$

The necessary conditions for optimal $\mathbf{w}(t)$ can be stated as follows.

THEOREM 3.1. *Consider the system given by* (14) *with the performance index given by* (15). *If* $\mathbf{w}(t)$ *minimizes* (15), *then there exists an adjoint vector* $\psi(t)$ *such that*

$$\frac{d\psi}{dt} = -\mathbf{A}^*\psi + (\mathbf{R}_1 - \lambda\mathbf{W}_1)\mathbf{x} + (\mathbf{R}_2 - \lambda\mathbf{W}_2)\mathbf{w}, \quad \psi(T) = 0, \quad (18)$$

and

$$\mathbf{w}(t) = (\mathbf{R}_3 - \lambda\mathbf{W}_3)^{-1}\{\mathbf{B}^*\psi - (\mathbf{R}_2 - \lambda\mathbf{W}_2)^*\mathbf{x}\}. \quad (19)$$

Proof. The theorem follows from the results of Chapters 1 and 2. To give a short proof, consider the alternate cost functional given by (17). By the maximal principle [12], the Hamiltonian is given by

$$H(\psi, \mathbf{x}, \mathbf{w}) = \psi^*(\mathbf{Ax} + \mathbf{Bw}) - \{\frac{1}{2}\mathbf{x}^*(\mathbf{R}_1 - \lambda\mathbf{W}_1)\mathbf{x}$$
$$+ \mathbf{x}^*(\mathbf{R}_2 - \lambda\mathbf{W}_2)\mathbf{w} + \frac{1}{2}\mathbf{w}^*(\mathbf{R}_3 - \lambda\mathbf{W}_3)\mathbf{w}\}. \quad (20)$$

The adjoint vector $\psi(t)$ satisfies

$$\frac{d\psi}{dt} = -\frac{\partial H}{\partial \mathbf{x}} \quad (21)$$

with the transversality condition $\psi(T) = 0$. Equation (18) is obtained from (21). Optimal $\mathbf{w}(t)$ is obtained by setting $\partial H/\partial \mathbf{w} = 0$ and is given by (19).   □

Let $\mathbf{V}_i = \mathbf{R}_i - \lambda\mathbf{W}_i$ for $i = 1, 2, 3$. We have a two-point boundary value problem given by

$$\begin{pmatrix} \dot{\mathbf{x}} \\ \dot{\psi} \end{pmatrix} = \begin{pmatrix} \mathbf{A} - \mathbf{BV}_3^{-1}\mathbf{V}_2^* & \mathbf{BV}_3^{-1}\mathbf{B}^* \\ \mathbf{V}_1 - \mathbf{V}_2\mathbf{V}_3^{-1}\mathbf{V}_2^* & -\mathbf{A}^* - \mathbf{V}_2\mathbf{V}_3^{-1}\mathbf{B}^* \end{pmatrix} \begin{pmatrix} \mathbf{x} \\ \psi \end{pmatrix}, \quad (22)$$

with

$$\mathbf{x}(t_0) = 0, \qquad \psi(T) = 0. \tag{23}$$

We now show that the minimum value of (15) is the least positive λ for which (22)-(23) has a solution with $\int_{t_0}^T \{\frac{1}{2}\mathbf{x}^*\mathbf{W}_1\mathbf{x} + \mathbf{x}^*\mathbf{W}_2\mathbf{w} + \frac{1}{2}\mathbf{w}^*\mathbf{W}_3\mathbf{w}\}\, dt > 0$.

THEOREM 3.2. *Consider the boundary value problem given by* (22) *and* (23). *Let* λ *be the least positive value for which the boundary value problem has a solution with* $\int_{t_0}^T \{\frac{1}{2}\mathbf{x}^*\mathbf{W}_1\mathbf{x} + \mathbf{x}^*\mathbf{W}_2\mathbf{w} + \frac{1}{2}\mathbf{w}^*\mathbf{W}_3\mathbf{w}\}\, dt > 0$, *where* $\mathbf{w}(t) = \mathbf{V}_3^{-1}\{\mathbf{B}^*\psi - \mathbf{V}_2^*\mathbf{x}\}$. *Then* λ *is the minimum value of* (15) *and* $\mathbf{w}$ *is an optimal input.*

Proof. From Theorem 3.1 it follows that if $\mathbf{w}(t)$ is optimal, then the boundary value problem (22)-(23) is satisfied for the optimal value of λ. Now suppose the boundary value problem is satisfied for some λ such that the corresponding solution $(\mathbf{x}, \psi)$ gives the denominator of (15) a positive value (with $\mathbf{w}(t) \overset{\triangle}{=} \mathbf{V}_3^{-1}\{\mathbf{B}^*\psi - \mathbf{V}_2^*\mathbf{x}\}$). We show that the performance index corresponding to $(\mathbf{x}, \psi)$ is λ.

Let $(\cdot, \cdot)$ denote the standard inner product in a real Euclidean space. We have

$$(\mathbf{R}_3 - \lambda\mathbf{W}_3)\mathbf{w} = \mathbf{B}^*\psi - (\mathbf{R}_2 - \lambda\mathbf{W}_2)^*\mathbf{x}.$$

Thus

$$\int_{t_0}^T \{(\mathbf{w}, \mathbf{R}_3\mathbf{w}) - \lambda(\mathbf{w}, \mathbf{W}_3\mathbf{w})\}\, dt = \int_{t_0}^T \{(\mathbf{w}, \mathbf{B}^*\psi) - (\mathbf{w}, \mathbf{R}_2^*\mathbf{x}) + \lambda(\mathbf{w}, \mathbf{W}_2^*\mathbf{x})\}\, dt. \tag{25}$$

Since $\mathbf{Bw} = \dot{\mathbf{x}} - \mathbf{Ax}$, the first integral on the right side of (25) is

$$\int_{t_0}^T (\mathbf{w}, \mathbf{B}^*\psi)\, dt = \int_{t_0}^T \{(\dot{\mathbf{x}}, \psi) - (\mathbf{Ax}, \psi)\}\, dt. \tag{26}$$

After integrating the right side of (26) by parts and utilizing $\mathbf{x}(t_0) = \psi(T) = 0$,

$$\int_{t_0}^T (\mathbf{w}, \mathbf{B}^*\psi)\, dt = \int_{t_0}^T \{(\mathbf{x}, \mathbf{R}_1\mathbf{x}) - (\mathbf{x}, \mathbf{R}_2\mathbf{w}) + \lambda(\mathbf{x}, \mathbf{W}_1\mathbf{x}) + \lambda(\mathbf{x}, \mathbf{W}_2\mathbf{w})\}\, dt. \tag{27}$$

Combining equations (25) and (27), we get

$$\int_{t_0}^{T} \{(\mathbf{x}, \mathbf{R_1 x}) + 2(\mathbf{x}, \mathbf{R_2 w}) + (\mathbf{w}, \mathbf{R_3 w})\} \, dt = \lambda \int_{t_0}^{T} \{(\mathbf{x}, \mathbf{W_1 x})$$
$$+2(\mathbf{x}, \mathbf{W_2 w}) + (\mathbf{w}, \mathbf{W_3 w})\} \, dt. \qquad (28)$$

Thus λ is the cost associated with $(\mathbf{x}, \psi)$. Thus, if λ is the least positive value for which the boundary value problem (22)-(23) has a solution $(\mathbf{x}, \psi)$ with the corresponding denominator of (15) being positive, then $\mathbf{x}$ must be an optimal trajectory.   □

If the system and weighting matrices are functions of a finite number of parameters, these parameters can be varied to maximize λ. In Section 2, since the system matrices and the weighting matrices depend on $\tilde{A}, \tilde{B}, \tilde{C}$, and $\tilde{D}$, an optimization routine needs to be employed with respect to these quantities to maximize λ.

4. OPTIMALITY CONDITIONS FOR THE MAXIMIZATION OF λ

We consider again the time-invariant H_∞ problem. In this section, we derive a condition that needs to be satisfied when λ is maximized. For this, consider equations (14) and (15). Note that for the standard H_∞ problem of Section 2, the system and weighting matrices depend on $\tilde{A}, \tilde{B}, \tilde{C}$, and $\tilde{D}$. These constitute the set of independent variables. The variations in the system and weighting matrices can be explicitly expressed in terms of variations in $\tilde{A}, \tilde{B}, \tilde{C}$, and $\tilde{D}$. However, the optimality conditions are extremely complicated to derive in such a case. The derivation can be simplified a little by assuming that $D_{22} = 0$ (see (3)). However, we only attempt to derive the basic optimality conditions here.

Consider equations (22) and (23). Let $\hat{A} = A - BV_3^{-1}V_2^*, \hat{B} = BV_3^{-1}B^*$, and $\hat{C} = V_1 - V_2 V_3^{-1} V_2^*$. Suppose $\tilde{A}, \tilde{B}, \tilde{C}$, and $\tilde{D}$ maximize λ. Let $\delta\tilde{A}, \delta\tilde{B}, \delta\tilde{C}$, and $\delta\tilde{D}$ denote elemental perturbations in $\tilde{A}, \tilde{B}, \tilde{C}$, and $\tilde{D}$ respectively. Also, denote the corresponding perturbations in $\hat{A}, \hat{B}, \hat{C}, \mathbf{x}, \psi$, and λ by $\delta\hat{A}, \delta\hat{B}, \delta\hat{C}, \mathbf{x}_1, \psi_1$, and μ respectively. Note that if λ is a maximum, $\mu = 0$. Thus, we have the following set of equations.

$$\dot{\mathbf{x}} = \hat{A}\mathbf{x} + \hat{B}\psi, \tag{29}$$

$$\dot{\psi} = \hat{C}\mathbf{x} - \hat{A}^*\psi, \tag{30}$$

$$\mathbf{x}(t_0) = \psi(T) = 0, \tag{31}$$

$$\dot{\mathbf{x}}_1 = \hat{A}\mathbf{x}_1 + \hat{B}\psi_1 + \delta\hat{A}\,\mathbf{x} + \delta\hat{B}\,\psi, \tag{32}$$

$$\dot{\psi}_1 = \hat{C}\mathbf{x}_1 - \hat{A}^*\psi_1 + \delta\hat{C}\,\mathbf{x} - \delta\hat{A}^*\psi, \tag{33}$$

$$\mathbf{x}_1(t_0) = \psi_1(T) = 0. \tag{34}$$

From (33), we have

$$\int_{t_0}^{T} \mathbf{x}^*\dot{\psi}_1 \, dt = \int_{t_0}^{T} \{\mathbf{x}^*\hat{C}\mathbf{x}_1 - \mathbf{x}^*\hat{A}^*\psi_1 + \mathbf{x}^*\delta\hat{C}\,\mathbf{x} - \mathbf{x}^*\delta\hat{A}\,\psi\} \, dt. \tag{35}$$

Also, by an integration by parts

$$\int_{t_0}^{T} \mathbf{x}^*\dot{\psi}_1 \, dt = -\int_{t_0}^{T} \{\mathbf{x}^*\hat{A}^*\psi_1 + \psi^*\hat{B}\psi_1\} \, dt. \tag{36}$$

From (35) and (36),

$$-\int_{t_0}^{T} \psi^*\hat{B}\psi_1 \, dt = \int_{t_0}^{T} \{\mathbf{x}^*\hat{C}\mathbf{x}_1 + \mathbf{x}^*\delta\hat{C}\,\mathbf{x} - \mathbf{x}^*\delta\hat{A}\,\psi\} \, dt. \tag{37}$$

From equation (30)

$$\hat{C}\mathbf{x} = \dot{\psi} + \hat{A}^*\psi. \tag{38}$$

Note that $\hat{C}^* = \hat{C}$. Substituting (38) in (37) and integrating by parts, we get

$$2 \int_{t_0}^{T} \mathbf{x}^* \delta \hat{A}\, \psi \, dt + \int_{t_0}^{T} \psi^* \delta \hat{B} \, \psi \, dt - \int_{t_0}^{T} \mathbf{x}^* \delta \hat{C} \, \mathbf{x} \, dt = 0. \tag{39}$$

The above equation needs to be satisfied for all elemental perturbations in $\tilde{A}, \tilde{B}, \tilde{C}$, and $\tilde{D}$.

5. A NUMERICAL EXAMPLE

As an example we consider the tracking problem given in [2]. The plant is given by

$$P(s) = \frac{s-1}{s(s-2)}. \tag{40}$$

The tracking error signal is $r - v$. The weighting filter $W(s)$ in Fig. 3 (p. 87) is given by

$$W(s) = \frac{s+1}{10s+1}. \tag{41}$$

The objective in [2] was to choose $K_1(s)$ and $K_2(s)$ such that the H_∞-norm of the transfer function from w to v is minimized. Our objective in this section is to synthesize u using the theory of this chapter such that the minimum of

$$\frac{\int_0^{10} w^2(t)\, dt}{\int_0^{10} \{(r-v)^2 + u^2\}\, dt} \tag{42}$$

is maximized.

Converting the plant equations to state space form, we have

$$\dot{x}_1 = -.1x_1 + w$$

$$\dot{x}_2 = u$$

$$\dot{x}_3 = 2x_3 + u \tag{43}$$

$$r = .1w + .09x_1$$

$$v = .5x_2 + .5x_3.$$

The matrices corresponding to equation (3) are given by

$$A = \begin{pmatrix} -.1 & 0 & 0 \\ 0 & 0 & 0 \\ 0 & 0 & 2 \end{pmatrix}, \quad B_1 = \begin{pmatrix} 1 \\ 0 \\ 0 \end{pmatrix}, \quad B_2 = \begin{pmatrix} 0 \\ 1 \\ 1 \end{pmatrix},$$

$$C_1 = \begin{pmatrix} .09 & -.5 & -.5 \\ 0 & 0 & 0 \end{pmatrix}, \quad C_2 = \begin{pmatrix} .09 & 0 & 0 \\ 0 & .5 & .5 \end{pmatrix},$$

$$D_{11} = \begin{pmatrix} .1 \\ 0 \end{pmatrix}, \quad D_{12} = \begin{pmatrix} 0 \\ 1 \end{pmatrix}, \quad D_{21} = \begin{pmatrix} .1 \\ 0 \end{pmatrix}, \quad D_{22} = \begin{pmatrix} 0 \\ 0 \end{pmatrix}.$$

The matrices F and H are chosen such that $A + B_2 F$ and $A + H C_2$ are stable. The choice is the same as that in [2] and is given by

$$F = (0 \quad .5 \quad -4.5), \quad H = \begin{pmatrix} 0 & 0 \\ 0 & 1 \\ 0 & -9 \end{pmatrix}.$$

Assume that $Q(s)$ is described by the three dimensional system

$$\dot{q} = \tilde{A}q + \tilde{B}\tilde{y},$$

$$u_2 = \tilde{C}q + \tilde{D}\tilde{y}. \tag{44}$$

Let $x = (x_1 \quad x_2 \quad x_3)^*$. Then the state equations for the finite-interval H_∞ problem become

$$\dot{x} = (A - B_2 \tilde{D} C_2)x + B_2(F + \tilde{D} C_2)\hat{x} - B_2 \tilde{C} q + (B_1 - B_2 \tilde{D} D_{21})w, \tag{45}$$

$$\dot{\hat{x}} = (A + H C_2 + B_2 F + B_2 \tilde{D} C_2)\hat{x} - (H C_2 + B_2 \tilde{D} C_2)x$$

$$-B_2 \tilde{C} q - (H D_{21} + B_2 \tilde{D} D_{21})w, \tag{46}$$

$$\dot{q} = \tilde{A}q + \tilde{B} C_2(x - \hat{x}) + \tilde{B} D_{21}w, \tag{47}$$

with the initial conditions being zero. The performance index is

$$\frac{\int_0^{10} w^2 \, dt}{\int_0^{10} \left\{ \begin{array}{l} (.1w + .09x_1 - .5x_2 - .5x_3)^2 \\ \quad + [F\hat{x} - (\tilde{C}q + \tilde{D}C_2 x - \tilde{D}C_2\hat{x} + \tilde{D}D_{21}w)]^2 \end{array} \right\} dt}. \tag{48}$$

Assuming values for $\tilde{A}, \tilde{B}, \tilde{C}, \tilde{D}$, we can find λ using the theory given in Section 3.

Let $\Phi = \begin{pmatrix} \Phi_{11} & \Phi_{12} \\ \Phi_{21} & \Phi_{22} \end{pmatrix}$ be the transition matrix corresponding to (22). Satisfaction of (23) gives rise to the condition that $\det(\Phi_{22}(10)) = 0$. Thus λ is found by making use of a sign change of $\det(\Phi_{22}(10))$ over a range of values of λ. In our numerical experiments, much of the computer execution time was consumed by the calculation of λ for a given controller. Efforts are under way to make the computation of λ more efficient.

The transition matrix $\Phi(10)$ was found in this case using the following formula [13]. Let $h = 10/2^8$. Represent the system matrix in (22) by $\mathcal{M}$. Then

$$\Phi(10) = \left\{ [I - \frac{1}{2}h\mathcal{M} + \frac{1}{12}h^2\mathcal{M}^2]^{-1} [I + \frac{1}{2}h\mathcal{M} + \frac{1}{12}h^2\mathcal{M}^2] \right\}^{2^8}. \tag{49}$$

Using the above procedure, we can iterate on $\tilde{A}, \tilde{B}, \tilde{C}$, and $\tilde{D}$ to maximize λ. Note that once $\Phi(h)$ is calculated, only eight repeated squarings are needed to evaluate $\{\Phi(h)\}^{2^8}$.

Initially the following values were assumed for the control matrices:

$$\tilde{A} = \begin{pmatrix} -2 & 0 & 0 \\ 0 & -2 & 0 \\ 0 & 0 & -2 \end{pmatrix}, \quad \tilde{B} = \begin{pmatrix} 1 & -1 \\ -1 & 1 \\ 1 & -1 \end{pmatrix}, \quad \tilde{C} = (1 \quad 1 \quad 1), \quad \tilde{D} = (1 \quad 1).$$

Using the Rosenbrock hill climbing algorithm [14], the elements of the matrices were varied to maximize λ. The algorithm usually leads to only local maxima. Note that $Q(s)$ is stable if and only if $\tilde{A}$ is stable. This was not introduced as a constraint in the optimization algorithm since the unconstrained run yielded a stable $\tilde{A}$. The Fortran program was run on a Zenith Z-248 personal computer in double precision using the Microsoft Optimizing Compiler Version 4.01. A local maximum of $\lambda = 14.8$ was obtained for the following values of $\tilde{A}, \tilde{B}, \tilde{C}$, and $\tilde{D}$:

$$\tilde{A} = \begin{pmatrix} -2.04 & .318 & .023 \\ -.026 & -1.632 & -.028 \\ -.052 & .358 & -2.054 \end{pmatrix}, \quad \tilde{B} = \begin{pmatrix} .945 & 9.973 \\ -1.046 & 33.028 \\ .946 & -1.056 \end{pmatrix},$$

$$\tilde{C} = (.944 \quad 1.48 \quad 1.018), \quad \tilde{D} = (.986 \quad 41.92).$$

After several runs with various initial values for $\tilde{A}, \tilde{B}, \tilde{C}$, and $\tilde{D}$, the value of $\lambda_{\max} = 14.8$ could not be bettered.

The two compoents of $Q(s)$ are given by

$$Q_1(s) = \frac{.986(s + 3.26)(s + 2.03)(s + .75)}{(s + 1.68)(s^2 + 4.05s + 4.1)},$$

$$Q_2(s) = \frac{41.92(s + 2.04)(s^2 + 5.04s + 6.58)}{(s + 1.68)(s^2 + 4.05s + 4.1)}.$$

(50)

It was reported in [2] that $Q_2(s)$ is unconstrained and may be taken as zero. To simulate this condition, we set the second columns of the optimal $\tilde{B}$ and $\tilde{D}$ equal to zero. The first positive value of λ for which $\det(\Phi_{22}(10))$ changed sign in this case was still observed to be 14.8.

A few comments on the numerical method are in order. Since the computation of λ consumes most of the execution time, further research needs to be done to find an alternate method to evaluate λ more accurately and efficiently. Also, the value of λ is evaluated in the above example by starting with an initial value and incrementing it in steps of 0.2 until a change in the sign of the determinant is observed. Thus the exact value of λ differs from the computed value by at most 0.2. This sort of inaccurate evaluation of λ may prematurely terminate the optimization routine which seeks to maximize λ.

6. CONCLUSIONS

A design methodology for the synthesis of finite-interval H_∞ controllers is presented using state-space methods. Using observer-based controller parametrization, an optimization problem is formulated. A measure of performance for a given controller is defined in terms of the least value of a parameter occurring in a two-point boundary value problem. Optimality conditions for finding the measure of performance for a given controller are given. The optimization problem seeks to maximize the measure of performance. An example is given.

Note: This chapter is based on a paper which will appear in the AIAA Journal of Guidance, Control, and Dynamics under the same title.

REFERENCES

[1] G. ZAMES, Feedback and optimal sensitivity: Model reference transformations, multiplicative seminorms, and approximate inverses, *IEEE Trans. Automat. Contr.*, Vol. AC-26, No. 2, 1981, 301-320.

[2] B. A. FRANCIS AND J. C. DOYLE, Linear control theory with an H_∞ optimality criterion, *SIAM J. Control Optim.* 25, 1987, pp. 815-844.

[3] B. A. FRANCIS, "A Course in H_∞ Optimal Control Theory," Lecture Notes in Control and Information Sciences, Vol. 88, Springer-Verlag, Berlin, New York, 1987.

[4] J. DOYLE, K. LENZ, AND A. PACKARD, "Design examples using μ-synthesis: Space shuttle lateral axis FCS during reentry," *Proc. IEEE Conf. on Decision and Control*, 1986.

[5] J. DOYLE, K. GLOVER, P. KHARGONEKAR, AND B. FRANCIS, "State-space solutions to standard H_2 and H_∞ control problems," Proc. Amer. Control Conf., 1988, pp. 1691-1696.

[6] K. GLOVER AND J. DOYLE, State space formulae for all stabilizing controllers that satisfy an H_∞ norm bound and relations to risk sensitivity," *Systems and Control Letters* 11, 1988, pp. 167-172.

[7] H. KIMURA AND R. KAWATANI, "Synthesis of H^∞ controllers based on conjugation," Proc. 27th IEEE Conf. Decision and Control, 1988, pp. 7-13.

[8] D. S. BERNSTEIN AND W. M. HADDAD, LQG control with an H_∞ performance bound: A Riccati equation approach, *IEEE Trans. Automat. Contr.* **34**, 1989, pp. 293-305.

[9] M. B. SUBRAHMANYAM, "Optimal disturbance rejection in time-varying linear systems," Proc. Amer. Control Conf., 1989, pp. 834-840.

[10] J. C. DOYLE, "Lecture Notes, ONR/Honeywell Workshop on Advances in Multivariable Control," Minneapolis, MN, 1984.

[11] B-C. CHANG AND A. YOUSUFF, " A Straight-Forward Proof for the Observer-Based Controller Parametrization," Proc. AIAA GNC Conf., Minneapolis, MN, 1988, pp. 226-231.

[12] E. B. LEE AND L. MARKUS, "Foundations of Optimal Control Theory," John Wiley, New York, 1967.

[13] L. LAPIDUS AND J. H. SEINFELD, "Numerical Solution of Ordinary Differential Equations," McGraw-Hill, New York, 1973.

[14] J. L. KUESTER AND J. H. MIZE, "Optimization Techniques with Fortran," McGraw-Hill, New York, 1973.

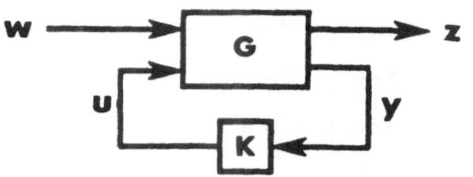

Fig. 1   The Standard Block Diagram

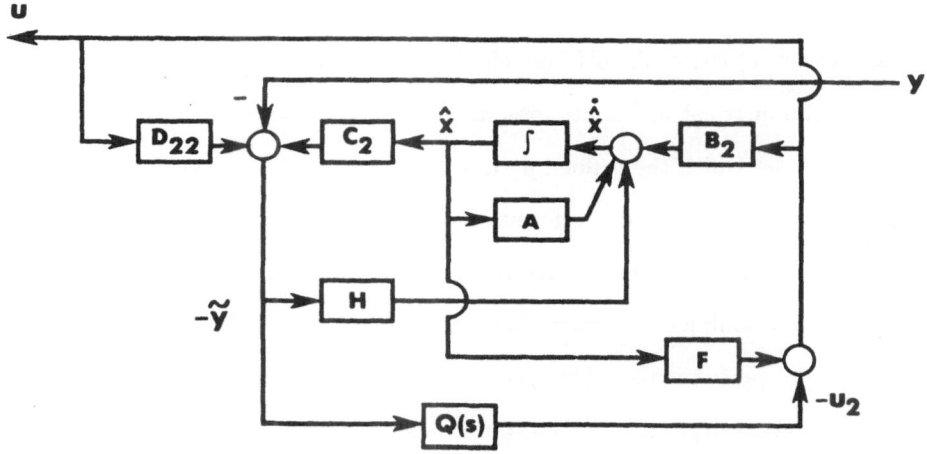

Fig. 2   The Observer-based Controller Parametrization

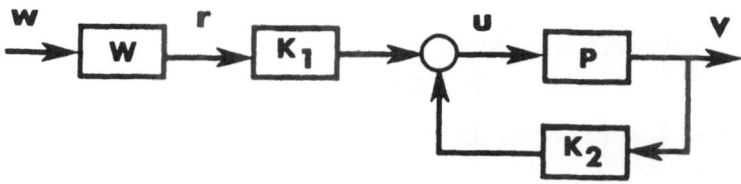

Fig. 3   Block Diagram for the Tracking Problem

CHAPTER 6

Worst-Case Performance Measures for

Linear Control Problems

ABSTRACT

In control systems, given a controller it is important to know the worst-case performance measure of the controller. In this chapter we define worst-case performance as the
minimum of a quotient of definite integrals. We give an existence result for the worst-case
conditions under which the defined performance measure is attained. Also we give a criterion for the evaluation of the performance measure by minimizing a parameter occurring
in a boundary value problem. Once the performance measure for a given controller can
be evaluated, a nonlinear programming algorithm can be used to choose a controller that
maximizes the performance. This problem can be considered as a generalization of the
H_∞-optimal control problem over a finite horizon. We consider convex integrands and
time-varying linear systems. We also give expressions for the variation of performance
owing to parameter and controller variations. These expressions are useful in evaluating
the robustness of the controller.

1. INTRODUCTION

In a control design problem the controller is selected to satisfy stability and performance requirements. In this chapter we concentrate on the aspect of performance of the
controller. Once a performance criterion is identified, it is important to know how to compute it given a stabilizing controller. This can aid in the selection of a controller which
maximizes the performance.

In this chapter we consider a general class of performance criteria. This class can be regarded as a generalization of the H_∞-optimality criterion and more details on this topic will be given later. Also results will be derived which aid in the computation of the performance for a given controller. Specifically, the problem treated in this chapter is given below.

Consider the time-varying linear system described by

$$\dot{x}_p = F(t)x_p(t) + B_1(t)v(t) + B_2(t)u(t), \qquad x_p(t_0) = 0, \tag{1}$$

$$\dot{x}_c = F_c(t)x_c(t) + B_c(t)y(t), \qquad x_c(t_0) = 0, \tag{2}$$

$$u = C_c(t)x_c(t) + D_c(t)y(t), \tag{3}$$

$$z = C_1(t)x_p(t) + D_1(t)u(t), \tag{4}$$

$$y = C_2(t)x_p(t) + D_2(t)u(t), \tag{5}$$

where $x_p(t), u(t), x_c(t), y(t), v(t)$ and $z(t)$ denote the state vector, the control vector, the control state vector, the output vector, the exogenous input vector, and the vector to be controlled respectively. The desired value of the error vector $z(t)$ is zero. We will relax the assumption that $x_p(t_0)$ and $x_c(t_0)$ be zero in Section 3.

We require that $I - D_c(t)D_2(t)$ be invertible for each t on the control interval $[t_0, T]$ so that the controller $u(t)$ can be expressed as

$$u(t) = \left(I - D_c(t)D_2(t)\right)^{-1}\left(C_c(t)x_c(t) + D_c(t)C_2(t)x_p(t)\right). \tag{6}$$

Assume for the present that the control matrices $F_c(t), B_c(t), C_c(t)$, and $D_c(t)$ are given. Let $v_0(t)$ be selected such that the quotient

$$\frac{\int_{t_0}^{T} \phi^1(v, t)\, dt}{\int_{t_0}^{T} \phi^2(z, t)\, dt} \tag{7}$$

attains a minimum value over all $v(t)$. For now we assume that $\phi^1(v,t)$ is convex in v for each $t \in [t_0, T]$. Also assume that both ϕ^1 and ϕ^2 are nonnegative. More assumptions will be imposed on these functions later. Let λ be the minimum value of (7). Thus $v_0(t)$ represents the worst exogenous input and λ gives a measure of the worst-case performance of the controller.

Once λ can be computed for given $F_c(t), B_c(t), C_c(t)$, and $D_c(t)$, these matrices need to be selected such that λ is maximized. For time-invariant controllers, this can be accomplished using a nonlinear programming algorithm. For time-varying controllers, these matrices need to be expressed in terms of basis functions and again a nonlinear programming algorithm needs to be used to maximize λ with respect to the coefficients of the basis functions.

If ϕ^1 and ϕ^2 are quadratic functions, (7) gives the ratio of the exogenous signal energy to the error energy. For time-invariant systems, the above problem then reduces to the H_∞-optimal control problem [1,2] provided that $T = \infty$. Our motive is to derive useful results in a general setting which subsumes the previous cases [3-5]. Ref. 3-5 consider the case in which ϕ^1 and ϕ^2 are quadratic functions. Theorems 2.1 and 3.1 of this chapter rely on the results developed in Ref. 6-11. The results of these papers were developed in a different context, namely, in the case of integral inequalities. We will not give complete proofs of Theorems 2.1 and 3.1 since these are basically given in Chapter 2.

We now give a summary of the results of this chapter. In Section 2, a result on the existence of the worst exogenous input for a given controller is obtained under certain assumptions. We assume in this case that the final time $T \leq \infty$. This result shows that the worst-case performance for the controller is actually attained under the assumptions given. In Section 3, it also forms a basis for the characterization of the performance measure of the worst exogenous input.

In Section 3 necessary conditions that are satisfied by the worst exogenous input are derived for a given controller. A two-point boundary value problem needs to be solved for the least positive value of the parameter λ to obtain the performance measure of the controller. A nonlinear programming algorithm needs to be used to find a controller which maximizes λ at least locally.

Many systems are subject to parameter variations and in Section 4, an expression is derived for the variation of performance of the controller as a functional of the variations in system parameters. This value gives an idea of the robustness of the controller and can aid in the choice of a controller with a specified level of robustness.

In Section 5 the topic of variation of performance owing to control parameters is treated. From the expressions in this section, the gradient of the worst-case performance with respect to control parameters can be found. Such a gradient evaluation can be especially useful in case the chosen nonlinear programming algorithm utilizes the gradient of the objective function.

Finally certain conclusions are given in Section 6.

2. EXISTENCE OF THE WORST EXOGENOUS INPUT

Assume that the matrices $F_c(t), B_c(t), C_c(t),$ and $D_c(t)$ are given. Writing (6) as

$$u(t) = M(t)x_p(t) + N(t)x_c(t), \tag{8}$$

we can rewrite equations (1)-(5) as

$$\begin{pmatrix} \dot{x}_p \\ \dot{x}_c \end{pmatrix} = \begin{pmatrix} F + B_2 M & B_2 N \\ B_c C_2 + B_c D_2 M & F_c + B_c D_2 N \end{pmatrix} \begin{pmatrix} x_p \\ x_c \end{pmatrix} + \begin{pmatrix} B_1 \\ 0 \end{pmatrix} v \tag{9}$$

with

$$z(t) = \begin{pmatrix} C_1 + D_1 M & D_1 N \end{pmatrix} \begin{pmatrix} x_p(t) \\ x_c(t) \end{pmatrix}, \tag{10}$$

where for brevity of notation, the dependence of the matrices on time has been omitted.

Let $x_p(t_0) = x_c(t_0) = 0$. Letting

$$x^* = (x_p^* \quad x_c^*)^*,$$

$$A(t) = \begin{pmatrix} F + B_2M & B_2N \\ B_cC_2 + B_cD_2M & F_c + B_cD_2N \end{pmatrix},$$

$$B(t) = \begin{pmatrix} B_1 \\ 0 \end{pmatrix},$$

and

$$C(t) = (C_1 + D_1M \quad D_1N),$$

we can write (9) and (10) as

$$\dot{x} = A(t)x + B(t)v, \quad x(t_0) = 0, \tag{11}$$

$$z(t) = C(t)x(t). \tag{12}$$

Note that the superscript * denotes matrix transpose.

Our problem in this section is to demonstrate the existence of a v for which the functional

$$F(z, v) = \frac{\int_{t_0}^T \phi^1(v(t), t)\, dt}{\int_{t_0}^T \phi^2(z(t), t)\, dt} \tag{13}$$

attains a minimum. This minimum value is a measure of the worst-case performance of the controller $u(t)$.

In this section the final time $T \leq \infty$. We make the following assumptions:

(i) $A(t), B(t)$, and $C(t)$ are continuous on $[t_0, T]$.

(ii) For i=1,2, ϕ^i is continuous in its arguments. Also ϕ^1 is convex in v for each t.

(iii) Admissible v are measurable functions such that $\int_{t_0}^T \phi^1(v, t)\, dt < \infty$.

(iv) $\phi^1(v,t) \geq a|v|^p, a > 0, p > 1$, and $\phi^2(z,t) \geq 0$ along any $z(t)$ which is the response to

some admissible $v(t)$.

(v) For each $K < \infty$, there is an integrable $g_K(t)$ such that if $\|v\|_p \leq K$, then

$$|\phi^2(z(t),t)| \leq g_K(t) \tag{14}$$

almost everywhere on $[t_0, T]$.

(vi) There exists $k > 0$ such that, for every $c \geq 0$,

$$\phi^1(cv,t) = c^k \phi^1(v,t),$$
$$\phi^2(cz,t) = c^k \phi^2(z,t). \tag{15}$$

(vii) There is an admissible v such that $0 < \int_{t_0}^T \phi^2(z,t)\,dt < \infty$.

The following proposition shows that the problem is equivalent to an isoperimetric problem.

PROPOSITION 2.1. *Consider* (11)-(13). *Let*

$$\inf_v \int_{t_0}^T \phi^1(v,t)\,dt = J \text{ subject to } \int_{t_0}^T \phi^2(z,t)\,dt = M > 0. \tag{16}$$

Also let

$$\lambda = \inf_v F(z,v) = \inf_v \frac{\int_{t_0}^T \phi^1(v,t)\,dt}{\int_{t_0}^T \phi^2(z,t)\,dt}. \tag{17}$$

Then $\lambda = J/M$.

Proof. Clearly $J/M \geq \lambda$. Now let $\tilde{v}$ be such that $F(\tilde{z}, \tilde{v}) \leq \lambda + \epsilon$ for some $\epsilon \geq 0$. Let $\int_{t_0}^T \phi^2(\tilde{z},t)\,dt = \tilde{M}$ and $\mu = (M/\tilde{M})^{1/k}$. Then $\int_{t_0}^T \phi^2(\mu\tilde{z},t)\,dt = M$ by assumption (vi) and $F(\mu\tilde{z}, \mu\tilde{v}) \leq \lambda + \epsilon$ by (15). Since ϵ is arbitrary, (17) follows.   □

THEOREM 2.1. *Consider* (11)-(13) *along with assumptions* (i)-(vii). *Then there exists an exogenous input for which* (13) *attains a minimum.*

Proof. By Proposition 2.1 it is enough to consider all exogenous inputs for which $\int_{t_0}^T \phi^2(z,t)\,dt = M > 0$. Let $\inf_v \int_{t_0}^T \phi^1(v,t)\,dt = J$ subject to $\int_{t_0}^T \phi^2(z,t)\,dt = M$. Let $\{v_n\}$ be such that $\lim_{n\to\infty} \int_{t_0}^T \phi^1(v_n,t)\,dt = J$ subject to $\int_{t_0}^T \phi^2(z_n,t)\,dt = M$. By assumption (iv), $\{v_n\} \subset L_p(t_0,T)$ and is bounded. Thus a subsequence, still denoted by $\{v_n\}$ converges weakly to some $v_0 \in L_p(t_0,T)$. Let z_0 be the response of (11) corresponding to v_0. By assumption (i) and by the weak convergence of $\{v_n\}$, $z_n(t) \to z_0(t)$ for all $t \in [t_0,T)$. By assumption (ii), $\phi^2(z_n,t)$ converges to $\phi^2(z_0,t)$ for all $t \in [t_0,T)$. Since $\{v_n\}$ is a bounded sequence in $L_p(t_0,T)$, by assumption (v) and by the Lebesgue dominated convergence theorem,

$$\int_{t_0}^T \phi^2(z_0(t),t)\,dt = \lim_{n\to\infty} \int_{t_0}^T \phi^2(z_n(t),t)\,dt = M.$$

Now following the reasoning in the proof of Theorem 2.1 of Chapter 2, it can be shown that

$$\int_{t_0}^T \phi^1(v_0,t)\,dt \le J.$$

Thus the proof is complete. □

3. CHARACTERIZATION OF THE OPTIMAL VALUE

In this section we derive a boundary value problem to be satisfied by an optimal exogenous input. This boundary value problem will have λ as a parameter. We will characterize λ as the least positive value for which the boundary value problem has a nontrivial solution. We consider again the system

$$\dot{x} = A(t)x + B(t)v, \qquad z = C(t)x, \tag{18}$$

with

$$x(t_0) = 0 \text{ or free}, \quad x(T) \text{ free}, \quad T < \infty, \tag{19}$$

$$F(z, v) = \frac{\int_{t_0}^{T} \phi^1(v, t)\, dt}{\int_{t_0}^{T} \phi^2(z, t)\, dt}. \tag{20}$$

We make the following assumptions.

(i) $A(t), B(t)$, and $C(t)$ are continuous on $[t_0, T]$.

(ii) $\int_{t_0}^{T} \phi^1(v, t)\, dt < \infty$ for any $v \in L_p(t_0, T), p > 1$.

(iii) $\phi^2(z(t), t) \geq 0$ almost everywhere on $[t_0, T]$ for any $z(t)$ which is the response of some

$v \in L_p(t_0, T)$.

(iv) Let ϕ^1 be continuously differentiable in v and ϕ^2 be continuously differentiable in x,

and both be measurable in t. Moreover, let $\phi_v^1 \in L_q(t_0, T)$ for all $v \in L_p(t_0, T), (1/p) +$

$(1/q) = 1$. Also, let ϕ_x^2 be bounded for bounded x, the bound being uniform for almost

all t.

(v) Let ϕ_v^1 and ϕ_x^2 be locally Lipschitzian in v and x respectively. For a definition of the

term locally Lipschitzian, see Section 3 of Chapter 2.

(vi) $v_0(t)$ minimizes (20) subject to (18) and (19)

$$\Rightarrow \quad 0 < \int_{t_0}^{T} \phi^1(v_0, t)\, dt < \infty, \qquad 0 < \int_{t_0}^{T} \phi^2(z_0, t)\, dt < \infty.$$

(vii) The pair $(A(t), B(t))$ is completely controllable.

We now state the necessary conditions for an optimal exogenous input.

THEOREM 3.1. *Consider the system (18)-(20). Suppose that $v_0(t)$ minimizes (20). Then*

there exists an adjoint response $\psi(t)$ such that

$$\frac{d\psi}{dt} = -A^*(t)\psi - \lambda\phi_x^2(z_0, t) \tag{21}$$

where

$$\lambda = \frac{\int_{t_0}^{T} \phi^1(v_0, t)\, dt}{\int_{t_0}^{T} \phi^2(z_0, t)\, dt} \tag{22}$$

and

$$\phi_v^1(v_0(t), t) - B^*(t)\psi(t) = 0 \quad \text{a.e. on } [t_0, T]. \tag{23}$$

Also

(a) if $x(t_0) = 0$ and $x(T)$ is free, then $\psi(T) = 0$;   (24a)

(b) if $x(t_0)$ and $x(T)$ are free, then $\psi(t_0) = \psi(T) = 0$.   (24b)

Proof. If $v_0(t)$ minimizes (20), then it also minimizes the alternate cost functional

$$\int_{t_0}^{T} \phi^1(v, t)\, dt - \lambda \int_{t_0}^{T} \phi^2(z, t)\, dt.$$

Equations (21)-(23) follow by formally applying the Pontryagin's maximum principle [12] to (18) with the alternate cost functional. A rigorous proof based on the Dubovitskii-Milyutin formulation [13] is given in Chapter 2 under assumptions (i)-(vii). Transversality conditions yield the boundary conditions given by (24).   □

Thus we have the following two-point boundary value problem given by

$$\dot{x} = A(t)x + B(t)v, \tag{25}$$

$$z(t) = C(t)x(t), \tag{26}$$

$$\phi_v^1(v, t) - B^*(t)\psi(t) = 0, \tag{27}$$

$$\dot{\psi} = -A^*(t)\psi - \lambda \phi_x^2(z, t), \tag{28}$$

with one of the following boundary conditions:

(a) $x(t_0) = 0$, $\psi(T) = 0$ (29a)

(b) $\psi(t_0) = 0$, $\psi(T) = 0$.   (29b)

We now give a characterization of λ as the minimum positive value for which the boundary value problem given by (25)-(29) has a solution with $\int_{t_0}^{T} \phi^2(z, t)\, dt > 0$.

THEOREM 3.2. *Let (x, v) be any pair that satisfies the boundary value problem given by (25)-(29) for some λ such that $\int_{t_0}^T \phi^2(z, t)\, dt > 0$. In addition to assumptions (i)-(vii), assume that there exists $k > 0$ such that, for every $c \geq 0$,*

$$\phi^1(cv, t) = c^k \phi^1(v, t),$$

$$\phi^2(cz, t) = c^k \phi^2(z, t). \tag{30}$$

Then

$$\frac{\int_{t_0}^T \phi^1(v, t)\, dt}{\int_{t_0}^T \phi^2(z, t)\, dt} = \lambda. \tag{31}$$

Proof. By (30) it follows that

$$v^* \phi_v^1 = k\phi^1(v, t)$$

$$z^* \phi_z^2 = k\phi^2(z, t). \tag{32}$$

From (28),

$$\int_{t_0}^T x^* \dot\psi\, dt = -\int_{t_0}^T x^* A^* \psi\, dt - \lambda \int_{t_0}^T x^* \phi_z^2\, dt. \tag{33}$$

Integrating the left side of (33) by parts and making use of (25) and (29), we get

$$\int_{t_0}^T x^* \dot\psi\, dt = -\int_{t_0}^T x^* A^* \psi\, dt - \int_{t_0}^T v^* B^* \psi\, dt. \tag{34}$$

From (27),

$$\int_{t_0}^T v^* B^* \psi\, dt = \int_{t_0}^T v^* \phi_v^1(v, t)\, dt. \tag{35}$$

From (33)-(35), we get

$$\int_{t_0}^T v^* \phi_v^1(v, t)\, dt = \lambda \int_{t_0}^T x^* \phi_z^2(z(t), t)\, dt. \tag{36}$$

Since $z(t) = C(t)x(t)$, $\phi_z^2 = C^* \phi_z^2$. From (32) and (36), we get

$$\frac{\int_{t_0}^T \phi^1(v, t)\, dt}{\int_{t_0}^T \phi^2(z, t)\, dt} = \lambda.$$

Thus the proof is complete.   □

Since the optimal $v(t)$ satisfies (25)-(29), we deduce from Theorem 3.2 that the minimum value of (20) is the same as the minimum positive λ such that (25)-(29) has a solution with $\int_{t_0}^T \phi^2(z,t)\,dt > 0$.

4. ROBUSTNESS CONSIDERATIONS

Many practical systems possess parameter uncertainties. In such cases it is desirable not to have excessive variation in performance owing to parameter variations. Thus, for a given controller, it is important to know the variation in λ with parameter variations. To be specific, assume that the controller is given and once again write (8)-(10) as

$$\dot{x} = A(t)x + B(t)v, \qquad z(t) = C(t)x(t), \tag{37}$$

with the matrices $A(t), B(t)$, and $C(t)$ as defined in Section 2. The initial condition is

$$x(t_0) = 0 \quad \text{or} \quad \text{free,} \tag{38}$$

and the performance index is

$$F(z,v) = \frac{\int_{t_0}^T \phi^1(v,t)\,dt}{\int_{t_0}^T \phi^2(z,t)\,dt}. \tag{39}$$

In addition to the assumptions (i)-(vii) of Section 3, we make the following assumptions.

(a) $\phi_{vv}^1(v,t)$ and $\phi_{zz}^2(z,t)$ exist and are continuous in v and z respectively, and both are measurable in t.

(b) There exists $k > 0$ such that for all $c \geq 0$

$$\phi^1(cv,t) = c^k \phi^1(v,t),$$

$$\phi^2(cz,t) = c^k \phi^2(z,t). \tag{40}$$

Let $\lambda = \inf_v F(z, v)$. Because of variations in the matrices of the original system (1)-(5), there will be corresponding variations in the matrices $A(t), B(t)$, and $C(t)$. Let the elemental dependent variations in A, B, and C be denoted by $\delta A, \delta B$, and δC respectively. For convenience, we will formulate the variation in performance in terms of $\delta A, \delta B$, and δC. Let μ be the variation in λ owing to $\delta A, \delta B$, and δC. Now the performance robustness problem can be stated as follows.

Performance robustness problem. Select a stabilizing controller such that λ is maximized with the side constraint $|\mu/\lambda| \le \mu_0$ for all $\|\delta A(t)\| \le a(t), \|\delta B(t)\| \le b(t)$, and $\|\delta C(t)\| \le c(t), \ t \in [t_0, T]$.

We now derive an expression for μ in terms of $\delta A(t), \delta B(t)$, and $\delta C(t)$. For almost all of the remainder of this section, we omit displaying the dependence of the functions on t for simplicity of notation.

For a given controller, let v minimize (39). From Section 3, we get the following two-point boundary value problem which needs to be satisfied by $x(t)$ and the adjoint vector $\psi(t)$.

$$\dot{x} = Ax + Bv \tag{41}$$

$$\phi_v^1 - B^* \psi = 0 \tag{42}$$

$$\dot{\psi} = -A^* \psi - \lambda \phi_x^2 \tag{43}$$

$$\text{Either } x(t_0) = \psi(T) = 0 \ \text{ or } \ \psi(t_0) = \psi(T) = 0. \tag{44}$$

Let x_1 and ψ_1 represent the variations in x and ψ due to $\delta A, \delta B$, and δC. Also, let v_1 be the corresponding variation in the exogenous input v that minimizes (39). Note that since $\phi_x^2 = C^* \phi_z^2$ and $\delta z = \delta C \, x + C x_1$,

$$\delta \phi_x^2 = \delta C^* \phi_z^2 + C^* \phi_{zz}^2 (\delta C \, x + C x_1). \tag{45}$$

From (41)-(43), we have the following equations that are satisfied by x_1, ψ_1, and v_1.

$$\dot{x}_1 = A x_1 + \delta A \, x + B v_1 + \delta B \, v \tag{46}$$

$$\phi_{vv}^1 v_1 - B^* \psi_1 - \delta B^* \psi = 0 \tag{47}$$

$$\dot{\psi}_1 = -A^* \psi_1 - \delta A^* \psi - \lambda \delta C^* \phi_z^2 - \lambda C^* \phi_{zz}^2 \delta C \, x - \lambda C^* \phi_{zz}^2 C x_1 - \mu \phi_z^2 \tag{48}$$

The boundary conditions are either

$$x_1(t_0) = \psi_1(T) = 0 \tag{49a}$$

or

$$\psi_1(t_0) = \psi_1(T) = 0. \tag{49b}$$

THEOREM 4.1. *Consider (41)-(44) and (46)-(49). Then the variation in λ is given by*

$$\mu = \frac{-\int_{t_0}^T \psi^* \delta A \, x \, dt - \int_{t_0}^T \psi^* \delta B \, v \, dt - \lambda \int_{t_0}^T \phi_z^{2*} \delta C \, x \, dt}{\int_{t_0}^T \phi^2 \, dt}. \tag{50}$$

Proof. From (48) we get

$$\int_{t_0}^T x^* \dot{\psi}_1 \, dt = - \int_{t_0}^T x^* A^* \psi_1 \, dt - \int_{t_0}^T x^* \delta A^* \psi \, dt - \lambda \int_{t_0}^T x^* \delta C^* \phi_z^2 \, dt$$

$$- \lambda \int_{t_0}^T x^* C^* \phi_{zz}^2 \delta C \, x \, dt - \lambda \int_{t_0}^T x^* C^* \phi_{zz}^2 C \, x_1 \, dt - \mu \int_{t_0}^T x^* \phi_z^2 \, dt. \tag{51}$$

Also, by an integration by parts and by (41), (44), and (49),

$$\int_{t_0}^T x^* \dot{\psi}_1 \, dt = - \int_{t_0}^T x^* A^* \psi_1 \, dt - \int_{t_0}^T v^* B^* \psi_1 \, dt. \tag{52}$$

From (51) and (52), we get

$$\mu \int_{t_0}^T x^* \phi_z^2 \, dt + \int_{t_0}^T x^* \delta A^* \psi \, dt + \lambda \int_{t_0}^T x^* C^* \phi_{zz}^2 C x_1 \, dt + \lambda \int_{t_0}^T x^* \delta C^* \phi_z^2 \, dt$$

$$+ \lambda \int_{t_0}^T x^* C^* \phi_{zz}^2 \delta C \, x \, dt = \int_{t_0}^T v^* B^* \psi_1 \, dt. \tag{53}$$

Note that by (40),

$$\int_{t_0}^{T} x^* \phi_x^2 \, dt = k \int_{t_0}^{T} \phi^2(z,t) \, dt. \tag{54}$$

From (47), we have

$$\int_{t_0}^{T} v^* B^* \psi_1 \, dt = \int_{t_0}^{T} v^* \phi_{vv}^1 v_1 \, dt - \int_{t_0}^{T} v^* \delta B^* \psi \, dt. \tag{55}$$

Also, from assumption (b)

$$x^* \phi_x^2(z,t) = z^* \phi_z^2(z,t) = k\phi^2(z,t) \tag{56}$$

and

$$x^* C^* \phi_{zz}^2 C = z^* \phi_{zz}^2 C = (k-1)\phi_z^{2^*} C = (k-1)\phi_z^{2^*}. \tag{57}$$

Making use of (57) and (43),

$$\lambda \int_{t_0}^{T} x^* C^* \phi_{zz}^2 C x_1 \, dt = -(k-1) \int_{t_0}^{T} (\dot\psi + A^* \psi)^* x_1 \, dt. \tag{58}$$

Integrating the first term of the integrand from the right side of (58) by parts and using (46), (44), and (49), we get

$$\lambda \int_{t_0}^{T} x^* C^* \phi_{zz}^2 C x_1 \, dt = (k-1)\left\{ \int_{t_0}^{T} \psi^* \delta A \, x \, dt + \int_{t_0}^{T} \psi^* B v_1 \, dt + \int_{t_0}^{T} \psi^* \delta B \, v \, dt \right\}. \tag{59}$$

Incorporating (54), (55), and (59) in (53), and utilizing the fact that

$$v^* \phi_{vv}^1 = (k-1)\phi_v^{1^*} = (k-1)\psi^* B, \tag{60}$$

and

$$x^* C^* \phi_{zz}^2 = z^* \phi_{zz}^2 = (k-1)\phi_z^{2^*}, \tag{61}$$

we get (50). Thus the proof is complete. □

Using (50), variation in the worst-case performance owing to parameter variations can be computed for any given controller.

5. VARIATION OF PERFORMANCE WITH CONTROL PARAMETERS

We need to use nonlinear pi gramming algorithms to maximize λ with respect to control parameters. Several nonlinear programming algorithms make use of the gradient of the objective function. Thus, it is useful to get an expression for the variation μ in λ owing to control variations.

For this, consider (37)-(39) with the same assumptions as in Section 4. From (8) and the definitions of matrices $A(t), B(t)$, and $C(t)$ in Section 2, it can be observed that only $A(t)$ and $C(t)$ are dependent on the controller. Let $\delta A(t)$ and $\delta C(t)$ be the variations in $A(t)$ and $C(t)$ respectively owing to variations in the controller matrices $F_c(t), B_c(t), C_c(t)$, and $D_c(t)$ (see (2) and (3)).

From (50), we get

$$\mu = \frac{-\int_{t_0}^{T} \psi^* \delta A\, x\, dt - \lambda \int_{t_0}^{T} \phi_z^{2^*} \delta C\, x\, dt}{\int_{t_0}^{T} \phi^2\, dt}. \tag{62}$$

Using (62) the gradient of λ with respect to control parameter variations can be computed.

From (62) a necessary condition that needs to be satisfied by the maximizing controller can be developed. Assume that the matrices $F_c(t), B_c(t), C_c(t)$, and $D_c(t)$ have a neighborhood in which the closed loop system maintains stability. Since $\mu = 0$ when λ is a maximum, it follows from (62) that the maximizing controller satisfies

$$\int_{t_0}^{T} \psi^* \delta A\, x\, dt + \lambda \int_{t_0}^{T} \phi_z^{2^*} \delta C\, x\, dt = 0. \tag{63}$$

6. CONCLUSIONS

In this chapter we extended the H_∞-optimality criterion to more general functionals. The notion of worst-case performance of the controller is defined for a given controller. An existence theorem is given which states the conditions under which the worst-case performance is actually attained. Also, a criterion is given which defines the worst-case performance as the least positive value of a parameter occurring in a two-point boundary value problem. When there are parameter variations, expressions are derived for the variation of the worst-case performance in terms of the parameter variations. These expressions are also useful in evaluating the gradient of the worst-case performance with respect to parameters characterizing the controller. They are especially meaningful in the design of an optimal controller in case the chosen nonlinear programming algorithm utilizes the gradient of the objective function. Finally, a necessary condition that needs to be satisfied by the optimizing controller is given.

REFERENCES

[1] B. A. FRANCIS, "A Course in H_∞ Optimal Control Theory", Lecture Notes in Control and Information Sciences, Vol. 88, Springer-Verlag, Berlin, New York, 1987.

[2] B. A. FRANCIS AND J. C. DOYLE, Linear control theory with an H_∞ optimality criterion, *SIAM J. Control Optim.* **25**, 1987, pp. 815-844.

[3] M. B. SUBRAHMANYAM, Synthesis of finite-interval H_∞ controllers by state space methods, *AIAA J. Guidance, Control, and Dynamics*, to appear.

[4] —————— , "Optimal disturbance rejection in time-varying linear systems," *Proc. Amer. Control Conf.* **1**, 1989, pp. 834-840.

[5] —————— , "Necessary conditions for the design of control systems with optimal disturbance rejection," *Proc. 28th IEEE Conf. Decision and Control*, 1989.

[6] —————— , Necessary conditions for minimum in problems with nonstandard cost functionals, *J. Math. Anal. Appl.* **60**, 1977, pp. 601-616.

[7] —————— , On applications of control theory to integral inequalities, *J. Math. Anal. Appl.* **77**, 1980, pp. 47-59.

[8] —————— , On applications of control theory to integral inequalities: II, *SIAM J. Control Optim.* **19**, 1981, pp. 479-489.

[9] —————— , A control problem with application to integral inequalities, *J. Math. Anal. Appl.* **81**, 1981, pp. 346-355.

[10] —————— , An extremal problem for convolution inequalities, *J. Math. Anal. Appl.* **87**, 1982, pp. 509-516.

[11] —————— , On integral inequalities associated with a linear operator equation, *Proc. Amer. Math. Soc.* **92**, 1984, pp. 342-346.

[12] E. B. LEE AND L. MARKUS, "Foundations of Optimal Control Theory," John Wiley, New York, 1967.

[13] I. V. GIRSANOV, "Lecture Notes in Economics and Mathematical Systems," No. 67, Springer-Verlag, New York, 1972.

Model Reduction with a Finite-Interval H_∞ Criterion

ABSTRACT

An important problem in flight control and flying qualities is the approximation of a complex high order system by a low order model. In this chapter, for a given reduced order model, we define the correlation measure between the plant and the model outputs to be the minimum of the ratio of weighted signal energy to weighted error energy. We give a criterion for the evaluation of the correlation measure in terms of minimization of a parameter occurring in a two-point boundary value problem. Once the correlation measure for a given reduced order model can be evaluated, a nonlinear programming algorithm can be used to select a model which maximizes the correlation between the plant and model outputs. The correlation index used can be regarded as an extension of the H_∞ performance criterion to the finite-interval time-varying case. However, the usual H_∞ problem seeks an optimal controller, whereas our problem is to select the reduced order model matrices which give the best correlation index. We also give an expression for the variation of the correlation owing to parameter variations and pose a robust model reduction problem. The utilization of the theory is demonstrated by means of some examples. In particular, a problem which involves the reduction of an unstable aircraft model with structural modes is worked out.

1. INTRODUCTION

Model reduction is an important problem in the case of airplanes with significant aeroservoelastic dynamics. The original model in such cases is of high order and thus, the resulting controller will have a complex structure, especially if it uses full state feedback. Also for highly augmented aircraft with flight and propulsive controls, it is useful to develop low order models to analyze flying qualities.

If the aim is to design a low order controller for a high order plant, there are at least three broad approaches to achieve this. A general account of these three approaches is given in [1]. The so called direct design methods assume a stabilizing controller of fixed degree and seek to find the controller that maximizes a quadratic performance index (see [2,3]).

Another approach is to get a high order controller by some design technique, such as LQG or H_∞, and then to approximate the high order controller by a low order one which possesses certain desirable properties. This approach is the subject matter of [1] and the pertaining literature is referenced in that paper.

The third approach is to approximate the high order plant by a low order one. Then a low order controller is designed and used to control the original plant. In this chapter we concentrate on this approach and consider the problem of approximating the original plant by a low order model in an optimal sense. This problem has been treated recently by several researchers under a variety of approximation criteria and we refer the reader to [4] for the relevant references. Although no computational results are given, [4] gives a sufficient condition which characterizes reduced order models satisfying an optimized L_2 bound as well as a prespecified H_∞ bound. The reduced order model is expressed in terms of solutions of four coupled algebraic Riccati equations.

We now state the main problem. For the sake of generality, we pose it for time-varying systems. Let the plant be described by

$$\dot{x}_p = A_p(t)x_p + B_p(t)u, \qquad x_p(t_0) = 0, \tag{1}$$

$$y_p = C_p(t)x_p + D_p(t)u, \tag{2}$$

where $x_p(t), u(t)$, and $y_p(t)$ denote the plant state vector, the control vector, and the plant output vector respectively. Let the reduced order model which approximates the plant be chosen to be

$$\dot{x}_m = A_m(t)x_m + B_m(t)u, \quad x_m(t_0) = 0, \tag{3}$$

$$y_m = C_m(t)x_m + D_m(t)u, \tag{4}$$

where $x_m(t)$ and $y_m(t)$ denote respectively the state vector and the output vector of the reduced order model.

For given $A_m(t), B_m(t), C_m(t)$, and $D_m(t)$, let u be chosen such that the correlation index given by

$$\frac{\int_{t_0}^{T} \frac{1}{2}u^*(t)R(t)u(t)\, dt}{\int_{t_0}^{T} \frac{1}{2}(y_p - y_m)^* Q(t)(y_p - y_m)\, dt} \tag{5}$$

is minimized. The superscript * denotes matrix or vector transpose. Let this minimum value be denoted by λ. Thus u represents the worst input and λ gives a measure of the worst-case correlation between the plant output and the model output. The problem is to choose $A_m(t), B_m(t), C_m(t)$, and $D_m(t)$ such that λ is maximized.

Since (5) represents the ratio of weighted signal energy to weighted error energy, the above problem may be regarded as a modified H_∞ problem except for a few differences. We consider time-varying systems and in our case the interval of control is finite. There are extensions of the H_∞ results to the finite-interval time-varying case [4]. However, our approach is different and is based on considering the inherent two-point boundary value problem. Also, the general aim of H_∞ problems is the design of an optimal controller, whereas in this chapter we are interested in the selection of model matrices. It is necessary in our case to use nonlinear programming algorithms in order to select the model matrices which maximize λ. In [5-7], we derived some results which aid in the selection of a controller which maximizes the worst-case performance. The results of [5] are presented in Chapter 5 and these will be utilized in Section 2 of this chapter.

In the case of time-invariant systems, a nonlinear programming algorithm can be used to find at least a local maximum of λ. For the time-varying case, the matrices $A_m(t), B_m(t), C_m(t)$, and $D_m(t)$ need to be expressed in terms of basis functions and a nonlinear programming algorithm needs to be used to maximize λ with respect to the coefficients of the basis functions.

We do not require the plant and the model to be open loop stable. This is significant since many of the modern aircraft have open loop unstable poles. We show in Section 4 by means of examples that the method is indeed applicable to such cases. There is yet another advantage of our method. One of the criticisms in the approach of getting a low order model from a high order plant is that the satisfactory approximation of the plant requires some knowledge in advance of the controller [1]. Since we maximize the correlation between the plant and model outputs for the worst possible input, the correlation in the case of any other controller is bound to be better. Thus, our method furnishes a satisfactory approximation without requiring an *a priori* knowledge of the controller.

We now give a summary of the results of the chapter. In Section 2, conditions that characterize the worst input are derived for a given model. A two-point boundary value problem needs to be solved for the least positive λ to obtain the worst-case correlation between the outputs of the plant and the model. A nonlinear programming algorithm can then be used to find the model matrices which maximize λ.

The stability and control derivatives of aircraft are subject to variations and it is also not possible to determine these exactly from wind tunnel data. There is already some interest in robust model reduction techniques [8]. In Section 3 we formulate a robust model reduction problem and derive an expression for the variation of correlation between the plant and model outputs as a functional of the variations in system parameters. This value gives an idea of the robustness of the approximate model and can aid in the choice of a reduced order model with a specified level of robustness.

In Section 4 some examples are worked out and details about the computational algorithm utilized are given. Correlation between the plant and the model is shown via time and frequency response plots. In order to keep the examples as simple as possible, we do not consider the robust model reduction problem in the case of these examples.

Finally, certain conclusions are given in Section 5.

2. COMPUTATION OF λ FOR A GIVEN REDUCED ORDER MODEL

Assume that the matrices $A_m(t), B_m(t), C_m(t)$, and $D_m(t)$ are given. In this section, we characterize λ as the minimum positive value for which a certain two-point boundary value problem has a nontrivial solution. Also, we derive a computationally useful characterization of λ.

Letting

$$x^* = (\, x_p^* \quad x_m^* \,)^*, \tag{6}$$

$$y = y_p - y_m, \tag{7}$$

$$A(t) = \begin{pmatrix} A_p & 0 \\ 0 & A_m \end{pmatrix}, \tag{8}$$

$$B(t) = \begin{pmatrix} B_p \\ B_m \end{pmatrix}, \tag{9}$$

$$C(t) = (\, C_p \quad -C_m \,), \tag{10}$$

and

$$D(t) = D_p - D_m, \tag{11}$$

we can write (1)-(4) as

$$\dot{x} = A(t)x + B(t)u, \quad x(t_0) = 0, \tag{12}$$

$$y = C(t)x + D(t)u. \tag{13}$$

The correlation index given by (5) can be put in the form

$$J(u) = \frac{\int_{t_0}^T \frac{1}{2}u^*(t)R(t)u(t)\,dt}{\int_{t_0}^T \{\frac{1}{2}x^*W_1x + x^*W_2u + \frac{1}{2}u^*W_3u\}\,dt}. \tag{14}$$

The problem is to characterize $u(t)$ that minimizes (14).

Let $\lambda = \inf_u J(u)$. We assume that for all t, $R(t) - \lambda W_3(t)$ is invertible, and

$$R(t) \geq 0, \tag{15}$$

$$\begin{pmatrix} W_1(t) & W_2(t) \\ W_2^*(t) & W_3(t) \end{pmatrix} \geq 0. \tag{16}$$

Equations (15) and (16) guarantee that the numerator and denominator of (14) are non-negative for any u. The necessary conditions that characterize the worst input can be stated as follows.

THEOREM 2.1. *Consider the system given by (12)-(14). If $u(t)$ minimizes the correlation index given by (14), then there exists an adjoint vector $\psi(t)$, not identically zero, such that*

$$\frac{d\psi}{dt} = -A^*\psi - \lambda W_1 x - \lambda W_2 u, \quad \psi(T) = 0, \tag{17}$$

with

$$u(t) = (R - \lambda W_3)^{-1}\{B^*\psi + \lambda W_2^* x\}. \tag{18}$$

Proof. For a proof, see Theorem 3.1 of Chapter 5.

Let

$$\hat{A} = A + \lambda B(R - \lambda W_3)^{-1} W_2^*, \tag{19}$$

$$\hat{B} = B(R - \lambda W_3)^{-1} B^*, \tag{20}$$

and

$$\hat{C} = -\lambda W_1 - \lambda^2 W_2 (R - \lambda W_3)^{-1} W_2^*. \tag{21}$$

Thus, we have a two-point boundary value problem given by

$$\begin{pmatrix} \dot{x} \\ \dot{\psi} \end{pmatrix} = \begin{pmatrix} \hat{A} & \hat{B} \\ \hat{C} & -\hat{A}^* \end{pmatrix} \begin{pmatrix} x \\ \psi \end{pmatrix} \tag{22}$$

with

$$x(t_0) = 0, \quad \psi(T) = 0. \tag{23}$$

The following theorem follows from Theorem 3.2 of Chapter 5.

THEOREM 2.2. *Let* (x, ψ) *satisfy the boundary value problem given by* (22) *and* (23) *for the least positive* λ *such that* $\int_{t_0}^{T} \{\frac{1}{2} x^* W_1 x + x^* W_2 u + \frac{1}{2} u^* W_3 u\} \, dt > 0$, *where* $u = (R - \lambda W_3)^{-1} \{B^* \psi + \lambda W_2 x\}$. *Then* λ *is the minimum value of the index given by* (14) *and* u *is the worst input.*

In [5] and [6], a computational technique which utilizes the transition matrix associated with (22) is given. This technique is also presented in Chapters 3 and 5 of this monograph. In this chapter we use the alternate technique given in Chapter 3. This technique is more stable numerically. The theory behind the technique is given below.

Let $\Phi(t, \tau)$ be the transition matrix associated with (22). Then we have

$$\begin{pmatrix} x(T) \\ \psi(T) \end{pmatrix} = \Phi(T, \frac{T + t_0}{2}) \Phi(\frac{T + t_0}{2}, t_0) \begin{pmatrix} x(t_0) \\ \psi(t_0) \end{pmatrix}. \tag{24}$$

Let

$$\Phi^{-1}(T, \frac{T+t_0}{2}) = \begin{pmatrix} \zeta_{11} & \zeta_{12} \\ \zeta_{21} & \zeta_{22} \end{pmatrix} \tag{25}$$

and

$$\Phi(\frac{T+t_0}{2}, t_0) = \begin{pmatrix} \nu_{11} & \nu_{12} \\ \nu_{21} & \nu_{22} \end{pmatrix}. \tag{26}$$

Multiplying (24) on the left by (25), we get

$$\begin{pmatrix} \zeta_{11} & \zeta_{12} \\ \zeta_{21} & \zeta_{22} \end{pmatrix} \begin{pmatrix} x(T) \\ \psi(T) \end{pmatrix} = \begin{pmatrix} \nu_{11} & \nu_{12} \\ \nu_{21} & \nu_{22} \end{pmatrix} \begin{pmatrix} x(t_0) \\ \psi(t_0) \end{pmatrix}. \tag{27}$$

Since $x(t_0) = \psi(T) = 0$,

$$\zeta_{11}x(T) = \nu_{12}\psi(t_0),$$

$$\zeta_{21}x(T) = \nu_{22}\psi(t_0). \tag{28}$$

Since the equations in (28) are linearly dependent, (28) has a nontrivial solution for $\psi(t_0)$ and $x(T)$ if and only if

$$\det \begin{pmatrix} \zeta_{11} & \nu_{12} \\ \zeta_{21} & \nu_{22} \end{pmatrix} = 0. \tag{29}$$

Thus, we can characterize λ as the least positive value for which (29) holds.

We can determine λ by doing a search over a range of positive values and picking the first value at which the determinant in (29) changes sign. We give more details on this in Section 4.

3. ROBUST MODEL REDUCTION

In this section we formulate a robust model reduction problem. The aim is to choose the best reduced order model under parameter variations. We derive an expression for the variation in the correlation measure λ in terms of variations in the system matrices. For simplicity of analysis, we assume that $D_p(t)$ and $D_m(t)$ in (11) are zero, which makes $D(t) = 0$.

Consider (1)-(10). The system equations are given by

$$\dot{x} = A(t)x + B(t)u, \quad x(t_0) = 0, \tag{30}$$

$$y = C(t)x. \tag{31}$$

We can write (5) as

$$\frac{\int_{t_0}^T \frac{1}{2} u^*(t) R(t) u(t)\, dt}{\int_{t_0}^T \frac{1}{2} x^*(t) C^*(t) Q(t) C(t) x(t)\, dt.} \tag{32}$$

For given $A_m(t), B_m(t)$, and $C_m(t)$, let λ be the minimum of the correlation index in (32) over $u(t)$. Let the elemental variations in $A_p(t), B_p(t)$, and $C_p(t)$ be denoted by $\delta A_p(t), \delta B_p(t)$, and $\delta C_p(t)$ respectively. Let $\delta A(t), \delta B(t)$, and $\delta C(t)$ be the variations in the matrices $A(t), B(t)$, and $C(t)$ corresponding to the elemental variations $\delta A_p(t), \delta B_p(t)$, and $\delta C_p(t)$. Notice that

$$\delta A(t) = \begin{pmatrix} \delta A_p(t) & 0 \\ 0 & 0 \end{pmatrix}, \tag{33}$$

$$\delta B(t) = \begin{pmatrix} \delta B_p(t) \\ 0 \end{pmatrix}, \tag{34}$$

and

$$\delta C(t) = \begin{pmatrix} \delta C_p(t) & 0 \end{pmatrix}. \tag{35}$$

Let μ denote the variation in λ caused by $\delta A, \delta B$, and δC. Now the robust model reduction problem can be stated as follows.

Robust model reduction problem. Find $A_m(t), B_m(t)$, and $C_m(t)$ such that

$$\inf_u \frac{\int_{t_0}^T \frac{1}{2} u^* R u\, dt}{\int_{t_0}^T \frac{1}{2} x^* C^* Q C x\, dt} \tag{36}$$

is maximized with the side constraint

$$|\mu/\lambda| \le \mu_0 \quad for\ all\ \ \|\delta A(t)\| \le a(t), \|\delta B(t)\| \le b(t), \ and\ \ \|\delta C(t)\| \le c(t), \tag{37}$$

where $a(t), b(t)$, and $c(t)$ are suitably chosen.

We now derive an expression for μ in terms of $\delta A(t), \delta B(t)$, and $\delta C(t)$. For given $A_m(t), B_m(t)$, and $C_m(t)$, let u minimize the index in (32). In the following, we suppress the dependence of the matrices on t for simplicity of notation. From (19)-(23), we get the following boundary value problem which needs to be satisfied by the corresponding pair (x, ψ).

$$\dot{x} = Ax + BR^{-1}B^*\psi, \tag{38}$$

$$\dot{\psi} = -\lambda C^*QCx - A^*\psi, \tag{39}$$

$$x(t_0) = \psi(T) = 0. \tag{40}$$

Let x_1 and ψ_1 represent the variations in x and ψ due to $\delta A, \delta B$, and δC. From (38)-(40), we have the following equations satisfied by x_1 and ψ_1.

$$\dot{x}_1 = Ax_1 + \delta A\, x + BR^{-1}B^*\psi_1 + (BR^{-1}\delta B^* + \delta B\ R^{-1}B^*)\psi, \tag{41}$$

$$\dot{\psi}_1 = -\mu C^*QCx - \lambda(\delta C^*QC + C^*Q\delta C)x - \lambda C^*QCx_1 - A^*\psi_1 - \delta A^*\psi, \tag{42}$$

$$x_1(t_0) = \psi_1(T) = 0. \tag{43}$$

Theorem 3.1. *Consider (38)-(43). Then the variation in λ is given by*

$$\mu = \frac{-\int_{t_0}^T \psi^*\delta A\, x\, dt - \int_{t_0}^T \psi^*B^*R^{-1}\delta B\, \psi\, dt - \lambda\int_{t_0}^T x^*C^*Q\delta C\, x\, dt}{\int_{t_0}^T \frac{1}{2}x^*C^*QCx\, dt}. \tag{44}$$

Proof. From (42), we get

$$\int_{t_0}^T x^*\dot{\psi}_1\, dt = -\mu\int_{t_0}^T x^*C^*QCx\, dt - \lambda\int_{t_0}^T x^*(\delta C^*QC + C^*Q\delta C)x\, dt$$

$$-\lambda\int_{t_0}^T x^*C^*QCx_1\, dt - \int_{t_0}^T x^*A^*\psi_1\, dt - \int_{t_0}^T x^*\delta A^*\psi\, dt. \tag{45}$$

Also, by an integration by parts and by (38), (40), and (43),

$$\int_{t_0}^T x\dot\psi_1 \, dt = -\int_{t_0}^T x^* A\psi_1 \, dt - \int_{t_0}^T \psi^* BR^{-1}B^*\psi_1 \, dt. \tag{46}$$

Since

$$\int_{t_0}^T x^*(\delta C^* QC + C^* Q\delta C)x \, dt = 2\int_{t_0}^T x^* C^* Q\delta C \, x \, dt, \tag{47}$$

from (45) and (46), we get

$$\mu \int_{t_0}^T x^* C^* QCx \, dt + 2\lambda \int_{t_0}^T x^* C^* Q\delta C \, x \, dt$$

$$+ \lambda \int_{t_0}^T x^* C^* QCx_1 \, dt + \int_{t_0}^T x^* \delta A^* \psi \, dt = \int_{t_0}^T \psi^* BR^{-1}B^*\psi_1 \, dt. \tag{48}$$

From (39),

$$\lambda \int_{t_0}^T x^* C^* QCx_1 \, dt = -\int_{t_0}^T (\dot\psi + A^*\psi)^* x_1 \, dt. \tag{49}$$

Integrating the first term of the integrand from the right side of (49) by parts, and using (40) and (43), we get

$$\lambda \int_{t_0}^T x^* C^* QCx_1 \, dt = \int_{t_0}^T \psi^* \delta A \, x \, dt + \int_{t_0}^T \psi^* BR^{-1}B^*\psi_1 \, dt$$

$$+ \int_{t_0}^T \psi^*(BR^{-1}\delta B^* + \delta B \, R^{-1}B^*)\psi \, dt. \tag{50}$$

Incorporating (50) in (48) and using the fact that

$$\int_{t_0}^T \psi^*(BR^{-1}\delta B^* + \delta B \, R^{-1}B^*)\psi \, dt = 2\int_{t_0}^T \psi^* BR^{-1}\delta B \, \psi \, dt, \tag{51}$$

we get (44).   □

Using (44), the variation in the correlation measure λ owing to parameter variations can be computed for any given $A_m(t), B_m(t)$, and $C_m(t)$.

4. NUMERICAL EXAMPLES

In this section we consider only time-invariant examples. The systems in the examples can be put in the form

$$\dot{x} = Ax + Bu, \quad x(0) = 0, \quad x^* = (\, x_p^* \quad x_m^* \,)^*, \tag{52}$$

with the correlation index

$$J(u) = \frac{\int_0^T \frac{1}{2} u^* R u \, dt}{\int_0^T \frac{1}{2} x^* C^* Q C x \, dt}, \tag{53}$$

where A, B, and C are given by (8)-(10). For given A_m, B_m, and C_m, let $\lambda = \inf_u J(u)$. To recap the procedure for finding λ, let

$$F = \begin{pmatrix} A & BR^{-1}B^* \\ -\lambda C^* QC & -A^* \end{pmatrix}, \tag{54}$$

and

$$\exp(\frac{FT}{2}) = \begin{pmatrix} \nu_{11} & \nu_{12} \\ \nu_{21} & \nu_{22} \end{pmatrix}, \tag{55}$$

$$\exp(\frac{-FT}{2}) = \begin{pmatrix} \zeta_{11} & \zeta_{12} \\ \zeta_{21} & \zeta_{22} \end{pmatrix}. \tag{56}$$

From (29), λ is given by the first positive value for which

$$\det \begin{pmatrix} \zeta_{11} & \nu_{12} \\ \zeta_{21} & \nu_{22} \end{pmatrix} = 0. \tag{57}$$

Now we iterate on the matrices A_m, B_m, and C_m using a nonlinear programming algorithm to maximize λ.

There are two primary computational algorithms that are needed to use this method of model reduction. They are a nonlinear global optimization routine to maximize λ and a relatively fast routine to compute λ. Currently λ is determined by finding the smallest positive value for which (57) holds. Due to the oscillatory nature of the value of the determinant as a function of λ, for suitable weighting matrices in (53), λ was originally calculated starting with an initial value of $\lambda = 0.1$ and incrementing by 0.2 until the determinant in (57) changed sign. While this method yielded accurate results, it also used excessive amounts of computational time.

To speed this process up, two modifications were made. First, λ was incremented by large steps until the value of the determinant was less than a percentage of its initial value. At this point, small increments in λ were applied until the determinant changed sign. This technique was successful in this case because the absolute value of the determinant never increased beyond a very small fraction of its initial value after the first zero crossing. The second modification was adaptively scaling down the input weighting matrix R so that the values of λ were consistently in the range of 10 to 20. This modification gives large enough values for accuracy and small enough values to decrease computational time.

The second necessary algorithm is a nonlinear global optimization routine. We are still in the process of developing an algorithm which satisfies both speed and accuracy requirements. For now, however, two methods were used to test our theory. For Example 1, we used a deterministic tunneling technique [9]. This technique in our case starts with a modified version of the Rosenbrock constrained hill climbing algorithm [10], searches for a better point than the current local maximum, and then restarts the hill climbing algorithm from there. While this method did converge to the global maximum, it also required an excessive number of iterations. The method used for Example 2 was a multi-start hill climbing algorithm with the starting point chosen by truncating the original system as well as by other model reduction techniques. This method will in general not converge to the global maximum without an excellent starting point, but it will find a good local maximum for a decent starting point.

The Rosenbrock hill climbing algorithm and the algorithm to compute λ were written in PC-MATLAB and run on a Zenith Z-248 personal computer.

Example 1. Simple illustrative examples

We will first show the reduction of a stable second order system and an unstable second order system. The system is of the form

$$\dot{x}_p = A_p x_p + B_p u, \qquad x_p(0) = 0, \tag{58}$$

$$y_p = C_p x_p, \tag{59}$$

where $A_p = \begin{pmatrix} 0 & 1 \\ -10 & -11 \end{pmatrix}$ for the stable case, and $A_p = \begin{pmatrix} 0 & 1 \\ 10 & -9 \end{pmatrix}$ for the unstable case. The other matrices are $B_p = \begin{pmatrix} 0 \\ 1 \end{pmatrix}$ and $C_p = (1 \quad 0)$.

Our first order model equations are

$$\dot{x}_m = a_m x_m + b_m u, \qquad x_m(0) = 0, \qquad y_m = c_m x_m. \tag{60}$$

The weighting matrices in (53) are $R = .01, Q = 1$, and the final time T is taken to be 2 seconds. Our optimization technique is the aforementioned tunneling algorithm.

In the stable case, the initial values were $a_m = -1, b_m = 1$, and $c_m = 1$. The value of λ increased from an initial value of 3.4 to a maximum of 16.8. The final reduced order model is given by $a_m = -.7485, b_m = .0944$, and $c_m = .9$. In the unstable case, the intial values were $a_m = 1, b_m = 1$, and $c_m = 1$ and the final values are given by $a_m = .9505, b_m = .0812$, and $c_m = 1.1119$. In this case, the value of λ increased from 2.8 to a final maximum of 17.4.

The time responses to a step input are given in Figs. 1 and 2 (p. 125). A comparison of the time responses shows excellent correlation between the plant and model outputs. The frequency responses are also well matched, at least up to 10 rad/sec. These can be seen in Figs. 3 and 4 (p. 126). Divergence in the frequency responses is to be expected, since no low order model can match the frequency response of the high order plant at sufficiently high frequencies.

Example 2. Aircraft with structural modes

In this example, an eighth order plant will be reduced into a fourth order system. The plant is the longitudinal system of the Advanced Supersonic Transport (AST) along with the two lowest frequency structural modes [11]. The structural modes are the first and second fuselage bending modes. The system is of the form

$$\dot{x}_p = A_p x_p + B_p u, \quad x_p(0) = 0, \tag{61}$$

$$y_p = C_p x_p, \tag{62}$$

$$x_p = (\, v \quad \alpha \quad \theta \quad q \quad x_1 \quad \dot{x}_1 \quad x_2 \quad \dot{x}_2 \,)^* \tag{63}$$

$$u = (\, \delta_e \quad \delta_t \quad \delta_c \quad \delta_a \,)^* \tag{64}$$

where the matrices A_p, B_p, and C_p are given in Table 1. In (63), the variables on the right side are, respectively, perturbed speed, angle of attack, pitch angle, pitch rate, first fuselage bending mode, its derivative, second fuselage bending mode, and its derivative. The quantities on the right side of (64) are the control inputs from the elevator, throttle, canard, and elevon, respectively. The flight condition is supersonic cruise at Mach 2.5. The units for the airspeed are ft/sec, and the angles and the control surface deflections are in degrees.

We attempted to reduce this to a fourth order system given by

$$\dot{x}_m = A_m x_m + B_m u, \quad x_m(0) = 0, \tag{65}$$

$$y_m = C_m x_m, \tag{66}$$

$$x_m = (\, v \quad \alpha \quad \theta \quad q \,)^*. \tag{67}$$

Table 1   Plant matrices of the AST longitudinal system

$$
A_p = \begin{pmatrix}
-0.0127 & -0.0136 & -0.0360 & 0.0000 & 0.0000 & 0.0000 & 0.0000 & 0.0000 \\
-0.0969 & -0.4010 & 0.0000 & 0.9610 & 19.5900 & -0.1185 & -9.2000 & -0.1326 \\
0.0000 & 0.0000 & 0.0000 & 1.0000 & 0.0000 & 0.0000 & 0.0000 & 0.0000 \\
-0.2290 & 1.7260 & 0.0000 & -0.7220 & -12.0200 & -0.3420 & 1.8420 & 0.8810 \\
0.0000 & 0.0000 & 0.0000 & 0.0000 & 0.0000 & 1.0000 & 0.0000 & 0.0000 \\
0.0000 & 0.1204 & 0.0000 & 0.0496 & -44.0000 & -1.2740 & -4.0300 & -0.5080 \\
0.0000 & 0.0000 & 0.0000 & 0.0000 & 0.0000 & 0.0000 & 0.0000 & 1.0000 \\
0.0000 & 0.1473 & 0.0000 & 0.3010 & -7.4900 & -0.1257 & -21.7000 & -0.8030
\end{pmatrix}
$$

$$
B_p = \begin{pmatrix}
0.0000 & 0.0194 & 0.0000 & 0.0000 \\
-0.0215 & 0.0000 & -0.0040 & -1.7860 \\
0.0000 & 0.0000 & 0.0000 & 0.0000 \\
-1.0970 & 0.0000 & 0.3660 & -0.0569 \\
0.0000 & 0.0000 & 0.0000 & 0.0000 \\
-0.6400 & 0.0000 & 0.1625 & -0.0370 \\
0.0000 & 0.0000 & 0.0000 & 0.0000 \\
-1.8820 & 0.0000 & 0.4720 & -0.0145
\end{pmatrix}
$$

$$
C_p = \begin{pmatrix}
1 & 0 & 0 & 0 & 0 & 0 & 0 & 0 \\
0 & 1 & 0 & 0 & 0 & 0 & 0 & 0 \\
0 & 0 & 1 & 0 & 0 & 0 & 0 & 0 \\
0 & 0 & 0 & 1 & 0 & 0 & 0 & 0
\end{pmatrix}
$$

Note that in this case, λ is a function of 48 independent variables and the matrix F in (54) is of dimension 24×24. The exponentials in (55) and (56) were evaluated using the built-in matrix exponential routine of PC-MATLAB. We observed that because of the large numbers involved, it is best not to invert the matrix in (55) to get the matrix in (56), but to compute it directly using the built-in routine. We ran the multi-start hill climbing algorithm with three different starting points, viz., with the truncated system matrices, and with the reduced order matrices from [11], which were obtained by balancing and spectral decomposition. The final time T was taken to be 5 sec, the weighting matrix Q was the identity matrix, and the weighting matrix R was selected as the diagonal matrix with all diagonal entries equal to 0.001. Although all the three runs yielded local maxima for λ, we obtained the best value of λ with the initial matrices obtained by spectral decomposition. In this case, the value of λ increased from 2.6 to 36.3. The resuting reduced order model is characterized by

$$A_m = \begin{pmatrix} -0.0112 & -0.0023 & -0.0258 & -0.0003 \\ -0.1248 & -0.4222 & 0.0139 & 0.8712 \\ -0.2817 & 0.0142 & 0.0152 & 1.0198 \\ -0.2862 & 1.7511 & -0.0032 & -0.6893 \end{pmatrix}, \tag{68}$$

$$B_m = \begin{pmatrix} 0.0027 & 0.0215 & 0.0013 & -0.0005 \\ 0.5117 & 0.0022 & -0.1354 & -1.8368 \\ -0.1422 & 0.0107 & 0.0449 & 0.0035 \\ -1.0341 & 0.0054 & 0.3664 & -0.0521 \end{pmatrix}, \tag{69}$$

$$C_m = \begin{pmatrix} 1.0427 & 0.0099 & 0.0035 & 0.0103 \\ -0.1416 & 1.0082 & 0.0028 & -0.0059 \\ -0.0047 & 0.0068 & 0.9932 & -0.0006 \\ -0.1852 & 0.0197 & 0.0128 & 1.0255 \end{pmatrix}. \tag{70}$$

The original unaugmented plant has the short period eigenvalues at 0.6687 (unstable), and -1.7755 (stable), the phugoid eigenvalues at $-0.0151 \pm i0.0886$, the first fuselage bending mode eigenvalues at $-0.7257 \pm i6.7017$, and the second fuselage bending mode eigenvalues at $-0.3122 \pm i4.4484$. The eigenvalues of A_m are given by $-0.0046, -0.0232, 0.7113$, and -1.7910. The correlation between the plant and model outputs is excellent in the chosen interval, and this can be seen from Figs. 5-8 (p. 127,128), where the responses to an elevator step input are plotted.

Even though this example has weakly coupled flexible and rigid body modes, it revealed another interesting feature. A comparison of the frequency responses in [11] using spectral decomposition and balancing demonstrated the superiority of the spectral decomposition method in this case. As a comparison, in Table 2, we list the initial and final values of λ with starting points obtained from the truncated system matrices, balancing, and spectral decomposition. It can be observed from Table 2 that spectral decomposition gives the best initial and final values for λ.

Table 2   Values of λ with different starting points

	Initial λ	Maximum λ
Truncation	0.1	2.0
Balancing	0.3	3.3
Spectral decomposition	2.6	36.3

While the starting point obtained from the spectral decomposition method gave the best value for the correlation measure, an improvement in the value of λ can be seen for all the starting points. This suggests another use of our method. Using our algorithm, we can fine-tune the reduced order models obtained by some other model reduction methods. We are currently developing an optimization algorithm utilizing certain characteristics of λ as a function of the reduced order model parameters. It is hoped that these characteristics will allow us to overcome the problems associated with the large number of local maxima and the sharp rise in the value of λ near the global maximum. Global optimization algorithms currently being considered include stochastic search methods [9], methods of global increase such as simulated annealing [9,12], and methods of improvement of local maxima such as tunneling [9].

As is mentioned at the beginning of this section, a vast amount of computation time was used up in the evaluation of λ for given reduced order model matrices. An important topic for further research is to devise an alternate method for the efficient evaluation of λ. Also, our evaluation of λ is only accurate to within the specified incremental step size of λ and this may lead to premature termination of the optimization routine that seeks to maximize λ.

5. CONCLUSIONS

In this chapter, we presented a technique for reduced order modeling with a modified H_∞ optimality criterion. A characterization for the determination of the correlation between plant and model outputs is given. Also the problem of robust model reduction is addressed and an expression for the variation of correlation with plant parameter variations is derived. Nonlinear programming algorithms were utilized to reduce the longitudinal flexible-body model of an Advanced Supersonic Transport. Further work needs to be done to devise a suitable global optimization algorithm.

REFERENCES

[1] B. D. ANDERSON AND L. Y. LIU, Controller reduction : Concepts and approaches, *IEEE Trans. Automat. Contr.* **34**, 1989, pp. 802-812.

[2] U. -L. LY, "A Design Algorithm for Robust Low Order Controller," Ph. D. dissertation, Department of Aeronautics and Astronautics, Stanford University, 1982.

[3] D. S. BERNSTEIN AND D. C. HYLAND, The optimal projection equations for fixed-order dynamic compensation, *IEEE Trans. Automat. Contr.* **29**, 1984, pp. 1034-1037.

[4] G. TADMOR, H_∞ in the time domain : The standard four block problem, *Mathematics of Control, Signals, and Systems*, to appear.

[5] M. B. SUBRAHMANYAM, Synthesis of finite-interval H_∞ controllers by state space methods," *AIAA Journal of Guidance, Control, Dynamics*, to appear.

[6] ——————— , "Optimal disturbance rejection in time-varying linear systems," *Proceedings of the American Control Conference*, Vol.1, 1989, pp. 834-840.

[7] ——————— , "Necessary conditions for the design of control systems with optimal

disturbance rejection," *Proceedings of the 28th IEEE Conference on Decision and Control,* 1989.

[8] W. M. HADDAD AND D. S. BERNSTEIN, Robust reduced-order modeling via the optimal projection equations with Petersen-Hollot bounds," *IEEE Trans. Automat. Contr.* **33**, 1988, pp. 692-695.

[9] A. H. G. RINNOOY KAN AND G. T. TIMMER, "Global Optimization : A Survey," *New Methods in Optimization and their Industrial Uses,* J. Penot (ed.), Birkhäuser Verlag, Basel, Boston, 1989, pp. 133-155.

[10] J. L. KUESTER AND J. H. MIZE, "Optimization Techniques with Fortran", McGraw-Hill, New York, 1973.

[11] R. D. COLGREN, "Methods for model reduction," *Proceedings of the AIAA Guidance, Navigation and Control Conference,* Part 2, 1988, pp. 777-790.

[12] S. KIRKPATRICK, C. D. GELATT, JR, and M. P. VECCHI, Optimization by simulated annealing," *Science* **220**, No. 4598, 1983, pp. 671-680.

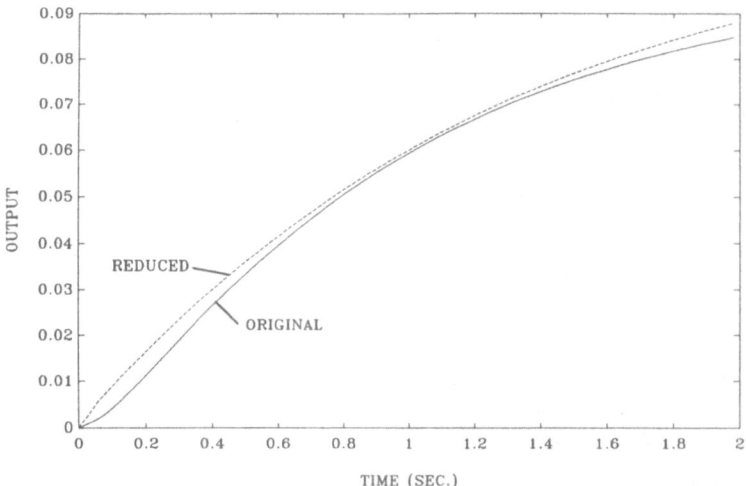

Fig. 1   Step Responses in the Simple Stable Case

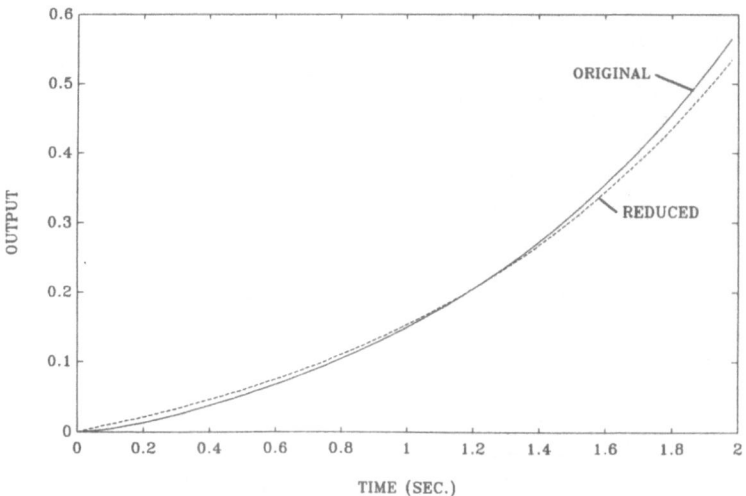

Fig. 2   Step Responses in the Simple Unstable Case

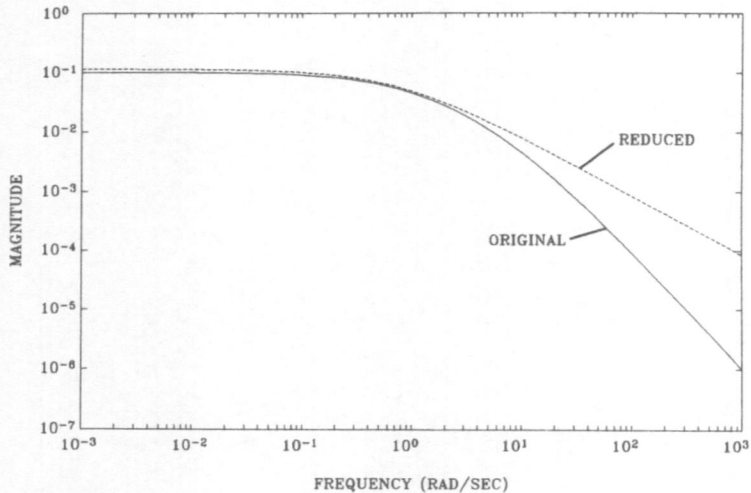

Fig. 3   Frequency Responses in the Simple Stable Case

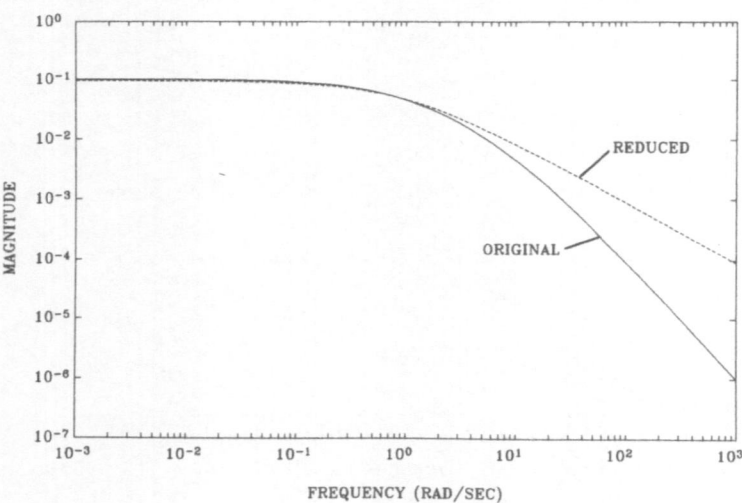

Fig. 4   Frequency Responses in the Simple Unstable Case

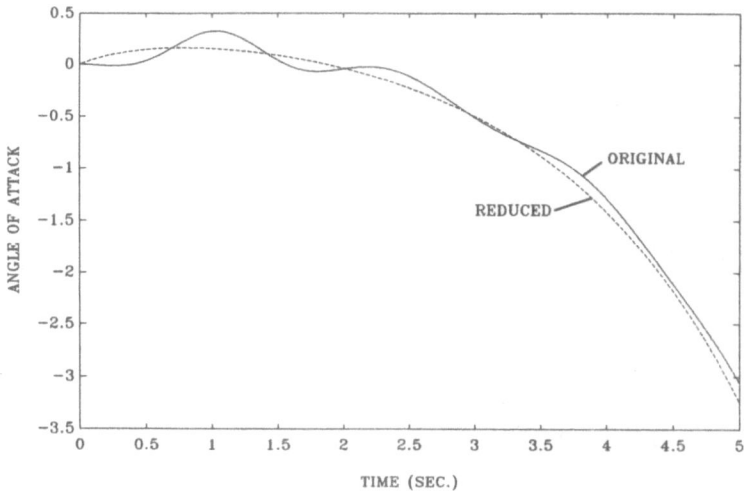

Fig. 5   Angle of Attack due to Application of an Elevator Step

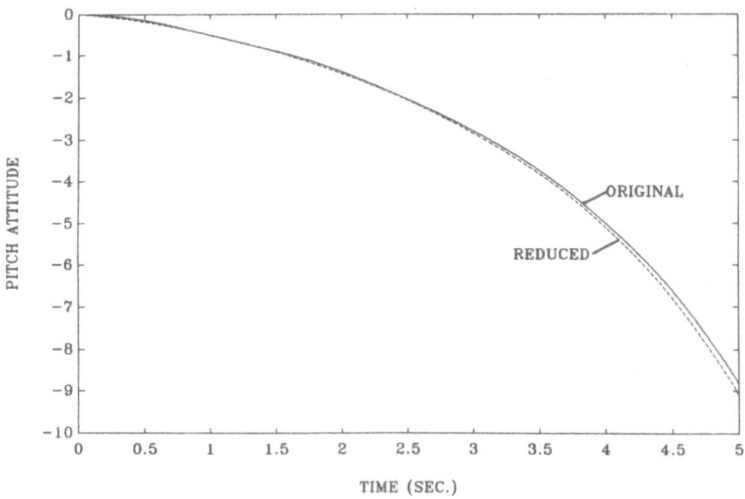

Fig. 6   Pitch Attitude due to Application of an Elevator Step

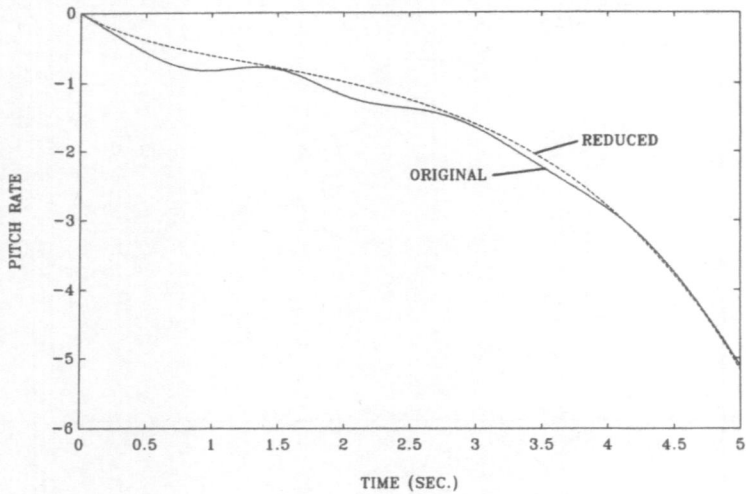

Fig. 7   Pitch Rate due to Application of an Elevator Step

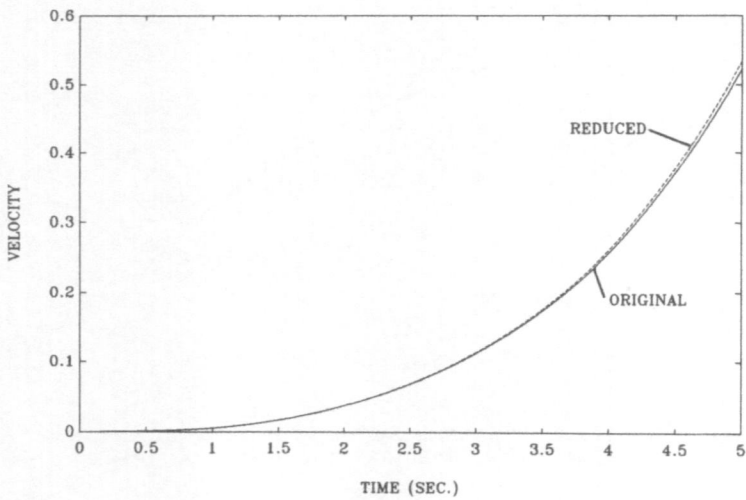

Fig. 8   Velocity due to Application of an Elevator Step

SUBJECT INDEX

Lecture Notes in Control and Information Sciences

Edited by M. Thoma and A. Wyner

Lecture Notes in Control and Information Sciences

Edited by M. Thoma and A. Wyner

Lecture Notes in Control and Information Sciences

Edited by M. Thoma and A. Wyner